BIODIGESTER
An Innovative Technology for on-board Disposal of Human Waste in Indian Railways

The Authors

Dr Lokendra Singh obtained his doctoral degree in Microbiology from Punjab Agricultural University and joined DRDO in 1984 at Defence Food Research Laboratory (DFRL) Mysore. He later joined as Scientist 'C' at Defence Laboratory (DL) Jodhpur and subsequently in 1989 at Defence Research and Development Establishment (DRDE) Gwalior. He has made significant contributions in the field of Environmental Microbiology and Biodefence. He participated in 14th Indian Scientific Expedition to Antarctica in 1993 which paved the way for the development of Biodigester Technology. Based on protracted laboratory experimentation and extensive field trials, he developed the Biodigester Technology suitable for all weather and terrain across India. These biodigesters are being utilized by Indian Railways as well as by various Government and non-Governmental organizations contributing immensely to 'Swachh Bharat Abhiyan'. He has also contributed significantly in the development of detection and identification systems including biosensor for toxins and pathogens of military and public health importance. He has 42 patents and more than 150 research publications in national and international journals. In 2009, Dr. Singh took over as Director, Defence Research Laboratory, Tezpur (Assam). He transformed the lab and led to development of various products. In 2012, he moved to DRDO HQ, New Delhi and served as Director, Directorate of Life Sciences and Director (Admin and Tech) in the same Directorate and coordinated the research activities of various laboratories of life sciences cluster. Dr Singh is also the recipient of several prestigious awards like DRDO Scientist of the year award in 2002, DRDO Technology Group Award in 2002, 2006, 2010 and 2012, Defence Technology Spin-off award in 2007, 2008 and 2014, and Life Time Achievement Award in Biotechnology and Pharmacy by the Association of Biotechnology and Pharmacy (ABAP) in the year 2016. Presently, Dr Lokendra Singh is holding the position of Outstanding Scientist and Director of Defence Research and Development Establishment (DRDE) at Gwalior.

Dr Dev Vrat Kamboj graduated from Haryana Agricultural University Hisar in Forestry Science. He obtained his master's and doctoral degrees from the same university in the field of Microbiology. He was recipient of Senior Research Fellowship (SRF) from Council of Scientific and Industrial Research (CSIR) for his Doctoral research work. Dr Kamboj joined DRDE Gwalior in 1998 as Scientist 'C' to work on the project on Human Waste Degradation. Dr Kamboj has contributed in the development and modification of designs related to variants of biodigesters for different regions across the country including Indian Railways. Dr Kamboj has 15 patents on the Biodigester Technology of which five have been granted in foreign countries. In addition, he has published around 55 research papers in journals of national and international repute. Dr Kamboj is a recipient of several prestigious awards like Young Scientist Award (1996) of Association of Microbiologists of India (AMI), DRDO Laboratory Scientist of the Year 2006 award, Technology Group Award (2002), Defence Spin-off Technology Award (2007, 2014) and DRDO Scientist of the year Award 2016 besides several best paper awards. He has represented India in several countries at various international fora. Dr Kamboj has also been reviewer for several prestigious international scientific journals. He is presently holding the position of Scientist 'F' and heading the Biotechnology Division.

BIODIGESTER

An Innovative Technology for on-board Disposal of Human Waste in Indian Railways

Lokendra Singh
Dev Vrat Kamboj

2018
Daya Publishing House®
A Division of
Astral International Pvt. Ltd.
New Delhi – 110 002

© 2018 AUTHORS

Cataloging in Publication Data--DK
Courtesy: D.K. Agencies (P) Ltd. <docinfo@dkagencies.com>

Singh, Lokendra, author.
Biodigester : an innovative technology for on-board disposal of human waste in Indian Railways / Lokendra Singh, Dev Vrat Kamboj.
  pages cm
  ISBN **9789387057531 (International Edition)**

  1. Railroads--Waste disposal--Technological innovations--India. 2. Sewage--Purification--Anaerobic treatment--India. 3. Indian Railways. I. Kamboj, Dev Vrat, author. II. Title.

  HE3296.S56 2018   DDC 625.150954   23

Published by   : **Daya Publishing House®**
                              A Division of
                              Astral International Pvt. Ltd.
                              – ISO 9001:2008 Certified Company –
                              4736/23, Ansari Road, Darya Ganj
                              New Delhi-110 002
                              Ph. 011-43549197, 23278134
                              E-mail: info@astralint.com
                              Website: www.astralint.com

Preface

To acquire a workable solution to a problem of Himalayan magnitude that remained unresolved, since 1909 for the Railways, is a sole satisfying endeavour. To design and establish an eco-friendly system for on-board degradation of human waste utilizing microbes for Indian Railways was a daunting task particularly when the operating parameters were quite wide and varied. Hardly any bacteria function effectively over a wide range of temperatures, nearly zero degree to 50 degrees Celsius.

The railways has emerged today as the main vehicle for socio-economic development of the country. However, the bane of Indian Railways has been the lack of sanitation. About 3980 MT of human waste is generated and dumped everyday by travelling passengers from 'Open discharge' module toilets of these coaches directly onto the rail tracks, polluting the stations and the areas through which the trains pass. This scenario has strong negative social and economic impact. Poor sanitation at railway stations has been a subject of ridicule from different quarters for quite some time.

To establish a biotechnology based sustainable system for the prevention of filth falling on railway tracks and stations is a *numero uno* effort. It is by seer perseverance and dogged determination of scientists of Defence Research & Development Organization (DRDO) who could find, propagate, modify and stabilize a bacterial culture and use it in the biodigester to achieve effective degradation of human waste. The degradation of faecal material is about 99 percent in the biodigester. The output that comes from the biodigester tank is only clear water and that can be utilized for different purposes like gardening, flushing etc.

World over, railway authorities have been trying to tackle the problem of poor sanitation caused by human excreta. This book presents glimpses of the effort undertaken by the Indian Railways in collaboration with DRDO to mitigate the problem of human waste disposal. Many of the foreign technologies that were tried in Indian scenario for on-board disposal of human waste were not found suitable, in addition to being costly. Many of them failed during initial runs also.

Indian Railways can take pride to have the most effective, economical and State-of-the-Art technology for on-board degradation of human waste by adopting *Biodigester Technology* developed by DRDO. As per available literature, currently this kind of technology is not being utilised elsewhere in the world. Even the latest generation of rail passenger cars, like the *French TGV*, the *Belgian Thalys* and *Amtrak's Acela* in the US, use lavatory units much like those found on passenger aircraft and act as a storage device that collects the human waste to be disposed elsewhere. None of these has the 'on-site' treatment mechanism.

Technological developments of biodigester with reference to Indian Railways have been presented chronologically in this book. Defence Research & Development Establishment (DRDE), a constituent laboratory of DRDO, had an in-depth knowledge of technology relating to biodigester and designing of biodigester tanks for different terrains and climatic conditions across the country as well as their installation. These biodigesters, however, had been operating in static mode, i.e., they were installed underground and remained as fixed fixtures. Designs of these biodigesters had freedom of space and options for construction materials. This knowledge became handy for Railways work.

Since the Biodigester Technology was functioning under static mode, and human-waste degradation system across India was to be tried in a dynamic mode of moving train, it was decided to go for protracted trials before adopting it for Indian Railways.

Designing biodigesters for Indian Railways had many constraints such as high input of night soil in the biodigester in short durations particularly in morning hours, limitation of space for the biodigester installation as well as fixing the biodigester tank underneath the passenger coach, and the material of construction as per the Railways norms. During train movement besides oscillations and jerks, the temperature variations during each journey as well as in different seasons were to be taken into account. One of the unforeseen problems that had to tackle was of finding solution to ward off dumping of non-degradable plastic materials in the biodigester inadvertently by passengers.

Surmounting numerous operational difficulties, not connected with the functioning of biodigester, the system proved functionally efficient. Along with the degradation efficiency, the system has withstood all fitment test parameters required during the train journey traversing through various temperature zones within the country.

The seamless embedding of an indigenously developed technology, for the on-board degradation of human waste in the coaches of passenger trains of Indian Railways, is set to bring neat and clean environment at stations and tracks. Clean environment at railway stations will reduce rodents, house flies, cockroaches and other insects infestation that feed on human excreta and help eliminate food related infections.

The new scenario will have a cascading effect on social and economic factors like sense of pride, sense of well being and clean surroundings which in turn will improve tourism. Due to its presence across the nation, Indian Railways could perhaps be the biggest contributor towards the objectives of 'Swachh Bharat Abhiyan'. Days are not far off when explicit as well as implicit benefits of this technology are extended to nations across globe and serve the mankind.

This book is an endeavour to compile the information about the Biodigester Technology developed for Indian Railways. Every effort has been made to keep the language of this book simple and lucid for the benefit of common man. The book has been arranged into seven chapters covering aspects on the impact of human waste on sanitation and diseases, efforts of Indian Railways for onboard disposal of human waste and the technological developments of biodigester during its adoption in Indian Railways, and finally future R&D prospects for improvement in the Biodigester Technology. It is believed that the book will create awareness about sanitation in society and facilitate the policy makers to adopt this technology for safe and scientific disposal of human waste in rural and urban societies in addition to Indian Railways.

Biodigester Technology has been very close to our heart owing to the association for a long time from conceiving the very idea to its successful implementation in Indian Railways. Authors are thankful to Dr RV Swamy, Dr K Sekhar and Dr R Vijayaraghavan, former Directors of DRDE Gwalior for their keen interest and hearty support for developing the technology. We are also highly grateful to Indian Railways authorities for accepting the technology and working with DRDO for adapting it for onboard treatment of human waste. Special thanks are due to Mr Sanjeev Kishore, the then Executive Director (ME/Dev) in Railway Board for realizing the potential of the technology and taking it forward in Indian Railways. This daunting task would not have been possible without profound enthusiasm of DRDE scientists in general and Team Biodigester in particular. When it came to compiling the information on this wonderful technology, the Team Biodigester once again contributed immensely, with equal zeal and enthusiasm, with their beautiful comments on the manuscript draft. Specifically, we express our sincere gratitude to Dr RS Chauhan, IDST Scientist, DRDE Gwalior for evincing his keen interest on the idea of writing the book, and for giving his valuable time with tirelessly sittings with us while shaping the compiled literature in the present book form, despite his own scientific commitments. We also wish to thankfully place on record the editorial assistance provided at various stages of this project by Dr MK Meghvansi, Scientist 'D' and Dr V Vasudevan, Scientist 'D' of the Team Biodigester. We also gratefully acknowledge the support provided by Mr Kanav Kansara and his team for M/s Astral International Pvt. Ltd., New Delhi for their keen interest in our proposal and timely support in bringing out this book within a very short span of time for our valuable readers.

Lokendra Singh
Dev Vrat Kamboj

Contents

1

Indian Railways and Sanitation

India is inhabited by one-sixth population of the earth that can take pride of its railway network. Indian Railways is one of the largest systems of modern transportation of the world. This single transport network connects far flung areas of the country and traverses its length and breadth providing the required connectivity and integration for balanced regional development. The system never rests; it has always been up and working unceasingly for the last several decades. Almost every citizen of the country is dependent on the railway network directly or indirectly. After US, Russian, and Chinese railways, India has the fourth-largest railway network in the world.

Indian Railways, being one of the largest transportation and logistics networks of the world, operates as many as 19000 trains both for passengers and goods. It runs 12000 trains to carry over 23 million passengers per day connecting about 8000 stations spread across the sub-continent. Indian Railways carried 8425.6 million passengers in 2013-14 which is about 1430 million more than the population of the world put together.

Indian Railways runs more than 7000 freight trains per day carrying about 3 million tonnes of goods every day. It has joined the select club of railways comprising Chinese, Russian and United States Railways with an originating freight loading of 1008.09 million tonnes (i.e., one billion plus) in 2012-13. During 2013-14, Indian Railways carried 1.05 billion tonnes of revenue earning freight traffic.

The growth on both carrying counts (Passenger and freight) has been very impressive, with the railways registering 160% growth over the last 15 years in freight and 200% in passengers. Besides being the cheapest way to move freight cargo, it also creates 60-80% lower carbon emissions than roads (for freight and passenger transport). It has the potential to become viable transportation alternative for the future generations that are bound to face the concerns of green and clean environment.

As an engine of growth and technology demonstration, it continues to build state-of-the-art carriages, high power locomotives and bridges across giant mountains and rivers. In an earlier era, due to the size and volume of work, the Indian Railways has been described as *"imperium in imperio"*, meaning an empire within an empire. The phrase has equal relevance in today's scenario also. The railways has emerged today as the main vehicle for socio-economic development of the country. Indian Railways has been instrumental in the development of human resources by way of imparting training on management of resources, design and development of a very large number of components, computer networking and communication, advanced signalling system and making of carriages.

Notwithstanding these praiseworthy feats, Indian Railways has been found wanting in the area of sanitation. About 3980 metric tonnes (MT) of human waste is generated and dumped everyday by travelling passengers from 'Open discharge' module toilets of these coaches that directly goes onto the rail tracks polluting stations and the areas through which the trains pass.

In the existing scenario, the discharge from toilets also falls directly into the water bodies or roads running below the railway bridges. The Indian Railways Bridge Manual emphasizes that rivers and roads under railway bridges should be covered by suitable and approved means to prevent droppings, falling from passing trains on water bodies or roads. Study by Indian Railways in 16 zones revealed that out of 1196, twenty per cent of road under-bridges were not covered at the bottom to prevent toilet discharge falling on the road users. Out of 5437 steel girder bridges across water bodies, about 70 per cent were not covered at the bottom leading to water bodies' pollution by toilet discharges falling from passing trains.

As a result of poor sanitation status, Indian Railways continued to be the target of all those who were concerned about the filth on tracks resulting in poor image of India abroad. Just to quote a few - In his book entitled ' An Area of Darkness', Nobel laureate Sir Vidiadhar Surajprasad Naipaul goes into a paroxysm of loathing about Indian hygiene: "Indians defecate everywhere. They defecate, mostly, beside the railway tracks".

Mr. Jairam Ramesh, erstwhile Cabinet Minister of Ministry of Drinking Water and Sanitation was so concerned about the sanitation that he went on saying that India is the world's capital of open defecation. "It is a matter of shame, anguish, sorrow and anger," he said. This shame will continue until all Indians have a toilet in their homes. Further he quotes Indian Railways as the 'largest toilet of the world'. Rose George in 2008 mentioned in her book entitled 'The Big Necessity' that of the two lakh tonnes of human faeces deposited daily in India, a large percentage is left on or alongside railway tracks.

It is rather unfair to isolate and blame Indian Railways alone for such poor state of sanitation. Indian Railways reflect the prevailing sanitation condition and attitude of masses in India. In spite of rich culture and heritage, India's record in following sanitation related discipline has been very poor. There are other contributing factors that erode the image of Indian Railways in terms of poor cleanliness. People occupying space along the Railway tracks not only use it for open defecation but also utilise the space for dumping all sort of waste including plastics and other non-degradable materials thus tarnishing the image of railways further. The onus falls on railways to clean their surroundings irrespective of the culprit being the third party.

Fig. 1.1: Garbage around the railway track

It has also been observed that human waste falling on the railway tracks tend to corrode the rails and make the structure weak. Replacing or maintaining the railway track has financial implications and cause inconvenience to smooth railway operations. Indian Railways spend on an average Rs. 3500 million every year for the manual scavenging, waste collection from tracks and resulting corrosion due to feacal material and contained moisture.

The other reason for hurling blame on Railways is the visibility of human waste generated filth that has far reaching consequences in terms of disease spreading. A gram of faeces can contain millions of viruses and bacteria, thousands of parasite cysts, and hundreds of worm eggs. One sanitation specialist has estimated that people who live in areas with inadequate sanitation ingest 10 grams of faecal matter every day. Therefore, it should not be surprising to accept by anyone that approximately 80 percent of the world's illness is caused by exposure to the faecal matter.

Since the presence of human excreta in water has been implicated in the transmission of many infectious diseases, like cholera, typhoid, hepatitis, polio, cryptosporidiosis and ascariasis, management of excreta and its proper disposal are of paramount importance. The profound disease burden attributed to diarrhoea and other diseases makes the human excreta as the most important target for the prevention of waterborne disease. Following respiratory illnesses, diarrhoeal diseases are the second leading cause of deaths from infectious disease, and the second leading cause of childhood mortality, exceeded only by neonatal conditions across the world.

As per the white paper on Indian Railways published by the Ministry of Railways in 2015, one of the biggest challenges being faced by Indian Railways is its inability to meet the demands of its customers for cleanliness. To address this issue, Indian Railways has been attempting to clean the filth on Railway tracks by resorting to various options like issuing instructions to passengers not to use toilets while train halts at stations (that is not followed by passengers as they find it more convenient to relieve themselves while train is stationary) as well as wet cleaning of filth at stations. Dislodging the faecal matter from the track by washing, only transfers waste from one place to another, finally making it to enter in to nearby water bodies. This has far reaching consequences in terms of spread of diseases. This contaminated water may find entry into nearby aquatic bodies or underground water pool that poses threat to the health of individuals residing in the vicinity.

Fig. 1.2: Cleaning of garbage by water from the railway track at the platform

The other option to clean the tracks is to engage manual scavengers. It is known that the Indian Railways is the biggest employer of manual scavengers. At times, the railways employees, individually and their unions, have raised the issue of stopping this inhuman practice. Government of India took cognizance of such practices prevalent throughout the country in both rural and urban sectors and enacted the Employment of Manual Scavengers and Construction of Dry Latrines (Prohibition) Act, 1993 prohibiting Manual Scavenging completely in the country. This Act serves as a primary instrument to eradicate the practice of manual scavenging. It is worth mentioning that there are serious health risks involved in manual handling of faecal matter.

The reason for the continuation of such a situation is perhaps the absence of suitable technology for treatment and disposal of human excreta. Toilet technology in the passenger trains all over the world has not improved much since its inception. The technology for the disposal of human excreta has moved a bit from 'Hopper Toilets' to variants of flushing mechanism with holding tank. The problem of excreta disposal on railway tracks has been a source of discomfort to passengers and rebuttal to railways throughout world.

Fig. 1.3: Removal of human excreta by manual scavenging

Perhaps the first patent design on railway toilet was filed in USA in 1915 (Patent number 1166291 A) to mitigate the problem to an extent. "The patent describes a method for the brakeman or other train staff to disable the flushing mechanism while the train is passing through a city. Once the flushing is enacted, the faecal matter is simply dumped onto the tracks".

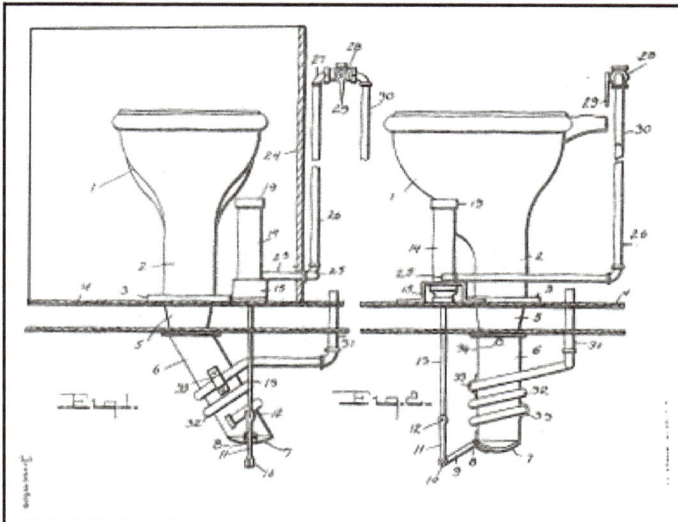

Fig. 1.4: Design of first hopper toilet developed in 1915 (USPTO 1166291 A)

The US patent 1325310 filed in 1918, describes a train toilet design with a holding tank. The holding tank partially filled with water collects the human waste from the toilet. When train reaches a speed of 40 miles per hour, the electrical solenoid energizes and allows the tank to drain onto the tracks.

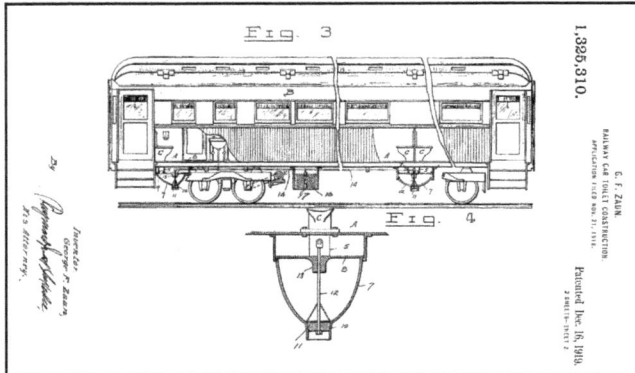

Fig. 1.5: Design of a railway coach toilet with provision for holding the excreta (USPTO 1325310)

Since around 1990 the US railways has been using holding tank system, while others only the drop chute toilets, i.e., 'Hopper Toilets'.

For Indian Railways the incorporation of 'in-carriage toilet' has an amusing beginning. A letter by Mr Okhil Chandra Sen to Indian Railways stirred the idea for the fitting of toilet in railway bogies. The letter proved to be an important one as according to the Railway Museum in New Delhi, the subsequent investigation into the affair by the British Raj resulted in the introduction of toilets to all trains in the country; something that had been absent since the formation of Indian Railways in 1857. The original letter is held in the museum's archives. Below is the version Indian Railway has on display at New Delhi Railway Museum.

Dear Sir,

      I am arrive by passenger train Ahmedpur station and my belly is too much swelling with jackfruit. I am therefore went to privy. Just I doing the nuisance that guard making whistle blow for train to go off and I am running with 'LOTAH' in one hand & 'DHOTI' in the next when I am fall over & expose all my shocking to man & female women on platform. I am got leaved Ahmed--pur station.

      This too much bad, if passenger go to make dung that dam guard not wait train minutes for him. I am therefore pray your honour to make big fine on that guard for public sake. Otherwise I am making big report to papers.

YOUR'S FAITHFULLY SERVENT,
OKHIL CH. SEN.

As per the literature, it has taken Railways a long time to move towards holding tank even in western world and inventors are still working to improve upon this problem. China, recently issued a patent for a toilet liquid sewage collection box for the train. While describing the current situation in China, the patent states that train sewage is 'directly discharged onto the railway roadbed' in China.

A March 2014 article "Mind The Crap: The continuing Problem of Poo on the Tracks" in *The Londonist* reported that all UK rolling stock built in 1996 has holding tanks, but older cars that dump the waste directly onto the tracks, including while in stations, were still in service. London's Kings Cross, Liverpool Street and Paddington stations are all served by trains with older cars and while passengers are urged not to use the toilets while at stations, it still happens. The track workers union was calling for train companies to add containment system, but while the government and the Rail Delivery Group is investing in new rolling stock, it was estimated that complete fleet upgradation would take several more years.

The holding tank design provides the opportunity for a cleaner station and the rail track. However, it requires holding tank to be emptied periodically. When the holding tank gets full the train crew may lock the lavatory shut because using the toilet in that condition may create mess inside the train instead of on the track.

The latest generation of rail passenger cars, like the French TGV, the Belgian Thalys and Amtrak's Acela in the US, use lavatory units much like those found on passenger aircraft.

These kind of systems are not suitable for adoption by the Indian Railways due to the large number of passengers using the toilet and the high volume of collected faecal matter produced due to wet ablution practice under Indian cultural conditions. Therefore, the holding capacity of tank has to be very high. Thus, the space constraint becomes the limiting factor for accommodation of such tanks beneath the bogie. Also the collected material has to be dumped somewhere, creating another problem at the place of disposal.

World over railway authorities have been trying to tackle the problem of poor sanitation caused by human excreta. As the passenger load on railways is increasing continuously, the problem of human waste disposal is becoming more severe.

Indian Railways has been attempting to tackle the problem of human waste disposal for quite some time and has accelerated its efforts since 1993. During the last 20 years or so, Indian Railways has approached IITs and other institutions for finding a suitable solution to mitigate the problem of filth falling on railway tracks from train-toilets. During 1993-95, Indian Railways in collaboration with Microphore, USA undertook trials of biotoilets in Grand Trunk Express running between New Delhi and Chennai. But these toilets had to be removed within six months due to clogging and foul smell. To address this problem, another trial with biotoilets modified by the Integral Coach Factory (ICF) was carried out which also could not solve the inherent problem of the treatment technology and could not come upto the expectation of Indian Railways.

Controlled Discharge Toilet System (CDTS) was also attempted by Indian Railways. After several years of trials in different trains, it was finally decided not to use the CDTS for disposal of human excreta generated onboard railway coaches as this system was also prone to chocking, not able to treat the excreta and rather dump the excreta on the track after the train attained a certain speed.

Around the same time frame, DRDO had developed a Biodigester Technology for onsite degradation of human waste. This technology had been tested at various geographical locations covering all extreme variants of temperature across India. After extensive trials, the technology emerged as sustainable, cost effective and easy to install. Further, these biodigesters could be fabricated by a range of materials. The degradation of faecal material was about 99% and the output being clear water was utilizable for gardening, flushing, etc. This technology had been in use at numerous places from Himalayas to Lakshadweep since the beginning of this century.

Once a technology is developed it takes time to cross technical, social-psychology and behavioural barriers for adopting in a different set up. This happened in the case of Indian Railways as well, with reference to Biodigester Technology. Since the Biodigester Technology, which was functioning as static mode human-waste degradation system across India, was to be tried in a dynamic mode of moving train, the Indian Railways opted for protracted trials before adopting it on pilot scale. Surmounting numerous operational difficulties, although not directly connected with the functioning of biodigester, the system proved to be efficiently functional. Along with the degradation efficiency, the system has withstood all fitment test parameters required during the train journey traversing through various temperature zones in the country.

Indian Railways can take pride in claiming to have the most effective, economical and State-of-the-Art technology for the on-board degradation of human waste after adopting the Biodigester Technology developed by DRDO. Since the Biodigester Technology is Green Technology and the resultant output from the tank is transparent harmless water devoid of any chemicals, it will go a long way in improving the cleanliness status of stations and the tracks throughout the length and breadth of the country. As per available literature, currently this kind of technology is not being utilised elsewhere in the world but has the potential to spread rapidly to other Railways.

Considering the vastness and the number of people utilising the railways, the sanitation message drawn from the cleaner rail tracks will go a long way in inculcating the cleanliness habits among masses. This will be a leap step forward towards national mission of 'Swachh Bharat Abhiyan' in making clean and green India.

2

Human Waste Associated Health Hazards

Microorganisms are the essential component of ecosystem and play a vital role for existence and survival of all sort of life on the planet earth. They are responsible for mineralization of essential elements and recycling of organic matter in different ecoecosystems starting from water to land, from sub-zero temperature of high altitude to scorching heat of deserts and hot water springs and from fresh water lakes to brackish waters of vast ocean. They also become almost the integral part of the gut of animal kingdom and help them not only for digestion of food and providing the essential vitamins but also prevent the opportunistic pathogens and other ingested pathogens from causing the diseases. Thus the microorganism, mainly the bacteria, act as friends and foe for the animals in day to day life, and human being are not exception to it. It is left to the wisdom of human being that he takes full advantage of these tiny creatures of God and adopt necessary measures to control them if they become the threat.

Gastrointestinal diseases have been attributed for death of millions of people annually mainly children and people of low income group especially of third world countries. Human waste being the main culprit transmits the pathogens from diseased persons to healthy individuals by contaminating food and drinking water sources. The masses are affected not only at personal level in terms of illness but also financially as it leads to inability to earn livelihood for them and for their family. It is well known that the poor health of the work force leads to collateral damage to the country's economy. The economic burden in India, due to human waste related sanitation amounted to Rs. 2.44 trillion in 2006 alone.

Human waste generally refers to human faeces and urine, the former being the by-product of digestion. It contains undigested material like cellulose and partly fermented unabsorbed materials like starch, fats and protein. Human waste also contains bile salts and microbial biomass originating from normal micro flora of the intestine besides pathogens if any. It is not essential that only those people who suffer from gastrointestinal diseases emits pathogens in their stool but a healthy person can also excrete pathogens in large quantities if she is a carrier without having any history of illness or after recovery from the particular disease.

Intestinal pathogens belong to bacteria (aerobic and anaerobic), viruses, protozoa and helminths (ova and cyst). According to a WHO report, a gram of stool (faeces) can contain up to 10 million viruses,1 million bacteria,100 parasite cysts and 100 worm eggs. The urine of a healthy individual is supposed to be sterile (free of microbes) but of unhealthy person may contain microorganisms of infectious diseases. It may contain pathogens like *Leptospira, Salmonella, Schistosoma,* and *E. Coli.* Thus human waste is rich source of pathogens which multiply with time as waste contains all the essential nutrients required by bacteria and other microbes to grow and thrive. Beside pathogens, human waste is a big aesthetic nuisance due to its physical appearance and bad odour.

Foul smell of the waste may be attributed to the presence of odorant volatiles like Methyl sulphides (methylmercaptan, dimethylsulfide & trisulfide), Benzopyrrole volatiles (indole, skatole) and Hydrogen sulfide. The odor of faeces may increase with various pathologies like Celiac disease, Ulcerative colitis, Chronic pancreatitis, Cystic fibrosis, Intestinal infection, e.g. *Clostridium difficile* infection, Malabsorption and Short bowel syndrome. Further, as the waste contains highly perishable constituents (partially digested) and huge number of bacteria, the decomposition continues in the environment producing more and more foul smell.

Human waste is composed of organic material along with water content and a little of inorganic matter consisting of mainly nitrogen, sulphur and phosphorus compounds in reduced form. Inorganic compounds get converted into oxidised or reduced forms by microbial activity and get released into the environment and are not of major concern till they get entry into the water bodies. Water gets evaporated in a day or more depending upon the humidity, temperature and wind velocity in the area. The moist organic waste may damage and reduce the life of railway tracks because of the chemical compounds present in the human waste. Further, microbial fermentation of the waste on the track generates acids, sulphides and H_2S gas that may further corrode the iron tracks. Replacing the railway lines is a costly affair and is a big financial burden on Indian Railways which is striving hard for the expansion of rail lines and trains besides improving the quality of services.

The human waste on tracks may be fermented by bacteria and/ or eaten by insects and rodents. The left over waste dries and gets mixed with the adjoining soil. In dry form the waste loses bad smell and pathogens to a significant level and thus the dehydrated waste is relatively of less concern. However, if the waste, in either form, gets entry into water bodies, it increases organic load of the water in terms of Biochemical / biological oxygen demand (BOD) and chemical oxygen demand (COD). High organic load causes the depletion of dissolved oxygen in water which is essential for survival of flora and fauna in water bodies like rivers, ponds and lakes. The microbes of water use the organic waste as substrate to meet their requirement of carbon, hydrogen and other elements and the energy for their survival and growth. During metabolism and respiration of bacteria, dissolved oxygen of water is also consumed. Aquatic animals like fishes are solely dependent for their survival on dissolved oxygen of the water which they consume through gills. Depletion of oxygen below five ppm leads to the death of such aquatic animals. Further dissolved oxygen is also required for the survival of aquatic flora. Hence contamination of organic matter in the water bodies leads to disturbance of aquatic ecosystem.

Besides organic matter, the pathogens of waste, even if they might have been reduced to lower numbers in dried waste, may again start multiplying in the contaminated water reaching to dangerous levels. Such water becomes the direct health risk to the consumers (human and animals) and also contaminates soil, fruits and vegetables.

The pathogens from human faeces thus get transmitted to the human population mostly by faecal–oral route or oral–faecal route that means pathogens in faecal particles passing from one host are introduced into the oral cavity of another host. The process of transmission of deadly microbes may be simple or involve multiple steps. Some examples of routes of faecal-oral transmission include:

- Potable water that has come into contact with the contaminated water
- Food that has been prepared with raw materials contaminated by faecal matter by any means
- Cooked food gets access to vectors like houseflies those spreading contamination from waste
- Cooked food exposed to contaminated dust
- poor or no cleaning of objects/ utensils that has been in contact with dust, flies or rodents
- not cleaning of hands and maintaining general hygiene

The "F-diagram" was first proposed in a publication by Hesperian Foundation for the United Nations Development Programme (UNDP) in 2005. It has been set up in a way that faecal-oral transmission pathways take place via nouns that start with the letter F: fingers, flies, fields, foods, and fluids (like polluted drinking water, surface water or groundwater).

House flies and rodents are two most important carriers of pathogens having direct relevance for transferring the contamination from human waste lying on railway track to the eatables and drinking water at railway stations.

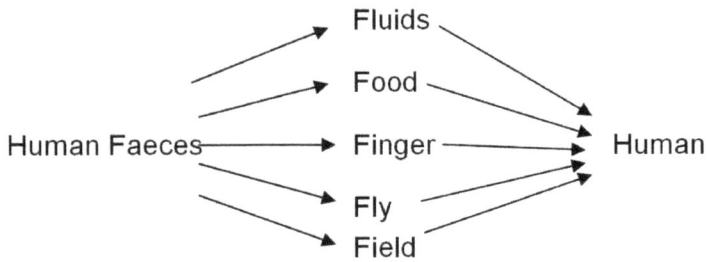

```
                        Fluids
                        Food
Human Faeces            Finger            Human
                        Fly
                        Field
```

House Flies

House flies are recognized as one of the most common carriers of communicable diseases. They feed and breed on human waste and other decomposing solid material. They are highly adaptable and can readily change their breeding habits. The diseases are caused by viruses, bacteria, protozoa and even cysts and ova of intestinal parasites. There are over 100 pathogens (disease-causing organisms) that are associated with house flies. Unlike pathogens which are transmitted by insects, such as mosquitoes or ticks, these pathogens do not specifically require an insect vector for transmission to human beings. The house fly plays no specific role in the life cycle of these pathogens. The fly is simply a carrier in some instances. House flies carry diseases on their legs and the small hairs that cover their bodies. It takes only a matter of seconds for them to transfer these pathogens to food or touched surfaces.

The fly is a restless insect and moves back and forth between food and filth. This helps in the spread of infection mechanically. As the fly vomits- frequently, the 'vomit drop' is often a culture of disease agents. Normally, houseflies remain close to their breeding places, but they disperse frequently up to 4 miles, and sometimes' even longer from the point of their origin.

Rodents

Railway stations have turned into a breeding ground for rats with packs of these rodents scurrying around the cloakroom, inside the parcel room, in the canteen and the restaurants. The rats on the tracks make burrows and look healthier than at other place indicating they have safe place to flourish as well as have abundant water and food to live. Rodents are often associated with infrastructural damage and eating/spoiling of stored food and products, however, their zoonotic risks are frequently underestimated. The best-known rat species are the black rat (*Rattus rattus*) and the brown rat (*Rattus norvegicus*). Wild rats can carry many different zoonotic pathogens, such as *Leptospira*, *Toxoplasma gondii*, and *Campylobacter*. The Black Death is traditionally believed to have been caused by the micro-organism *Yersinia pestis* which is well known causative agent of plague and is carried by the tropical rat flea (*Xenopsylla cheopis*). Besides, wild rats transmit foul smelling organic matter and whatsoever bacteria present in waste to eatables and other goods.

Some of the important diseases caused by human faeces are described below:

Cholera

Vibrio cholerae is a highly motile Gram-negative bacterium which can grow in the presence as well as absence of oxygen. Morphologically they are slightly curved, however can also appear as straight rods. All the strains of *V. cholera* do not cause cholera and only a few have been found responsible for the disease. The pathogenic strain should colonize in the small intestine via the toxin co regulated pilus (a thin, flexible, filamentous appendage) on the surface of bacterial cell and produce cholera toxin (a protein) for causing the actual illness.

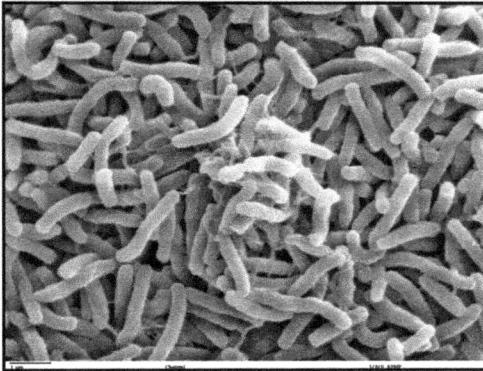

Fig. 2.1: *Vibrio cholerae*

(Source: https://upload.wikimedia.org/wikipedia/commons/9/9d/Cholera_bacteria_SEM.jpg)

Cholera is one of the most devastating human diseases, with a rapid onset of diarrhoea (characteristic rice water stool), abdominal cramps and vomiting within few hours to 2-3 days of ingestion. Massive fluid losses (15-20 litres/day) leads to dehydration with symptoms of thirst, dry mucous membranes, decreased skin turgor, sunken eyes, hypotension, weak or absent radial pulse, tachycardia, cramps, renal failure, coma, and death. The disease typically lasts for 4–6 days. In case of untreated individuals, the death can occur in as many as sixty percent of cases, however, rehydration treatment can prevent the death.

The main reservoirs of *V. cholerae* are people and aquatic sources such as brackish or salt water including its flora and fauna. Cholera infections are most commonly acquired from drinking water in which *V. cholerae* has been introduced from the faeces of an infected person. The surface water is easily accessible to be contaminated with human faeces in economically strained countries where open defecation is very common and even if toilets are used, waste treatment is of low priority. Shallow hand pumps get contaminated with the bacteria as it can easily travel through the layers of thin soil but in case of sandy soil even deep water pumps are prone for contamination as coarse sandy soil allows the bacteria to traverse longer distances.

Other common vehicles of contamination include fish, vegetables and grains grown with polluted water especially if they are eaten raw or not heated properly during cooking. Transmission of disease, from person to person, through contact is not possible as required number of bacteria for causing the infection is very high.

Two serogroups of *V. cholerae*, O1 and O139, have been reported to cause outbreaks of cholera, worldwide. Serogroup O1 has been responsible for the majority of outbreaks, while O139 (first identified in Bangladesh in 1992) is confined to Southeast Asia. Many other serogroups of *V. cholerae*, with or without the cholera toxin gene can also cause diarrhoeal disease similar to cholera.

Shigellosis

Shigellosis is also known as the bacillary dysentery. *Shigella* is a Gram negative rod shape bacterium that causes dysentery due to the infection and destruction of epithelial cells of the intestinal mucosa in the caecum (last part of small intestine) and rectum. Dysentery can be caused by as few as 100 bacterial cells. It is estimated that *Shigella* causes approximately 90 million cases of severe dysentery annually, with at least 100,000 of these resulting in death, mostly among children in the developing world. It is common in human beings and is hardly reported in animals.

This causative organism is also often found in water polluted with human faeces, and is transmitted via faecal-oral route. *Shigella* can be transmitted through food, including salads, raw vegetables, milk and other dairy products, and meat contaminated with faecal matter. Water contaminated with faecal matter and unhygienic handling of food by food handlers are the most common causes of shigella dysentery. Bacteria is also transmitted through fomites (materials like clothes, utensils, furniture, etc), water and mechanical vectors like houseflies besides hand-to-mouth infection.

Signs and symptoms may range from mild abdominal discomfort to full-blown dysentery characterized by cramps, diarrhoea, with blood and mucus-consistent stools and also fever. The onset of shigella dysentery takes 12 to 96 hours after infection, and the recovery usually takes 5 to 7 days. Some strains of *Shigella dysenteriae* produce the enterotoxin, shiga toxin, which is a cellular protein inhibitor and associated with hemolytic uremic syndrome that leads to fatal kidney failure.

Salmonellosis

Salmonellosis *is caused by* infection of a Gram negative rod shaped bacterium called *Salmonella*. The disease usually occurs when a person eats food contaminated with the faeces of animals or humans carrying the bacteria. *Salmonella* outbreaks are commonly associated with eggs, meat and poultry, but these bacteria can also contaminate other foods such as fruits and vegetables.

Salmonella is generally divided into two categories. Non-typhoidal *Salmonella* (*Salmonella typhimurium*) is the most common form, and is carried by both humans and animals. Typhoidal *Salmonella*, which causes typhoid fever, is rare, and is caused by *Salmonella typhi*, which is carried only by humans.

Symptoms of *Salmonella* infection range widely, and are sometimes absent altogether. The most common symptoms include diarrhoea, abdominal cramps, and fever (100-102 °F). The person may also suffer with bloody diarrhoea, vomiting, headache and body ache. Typical symptoms appear in 6 to 72 hours after eating the contaminated food and may last for 3 to 7 days without treatment. Symptoms of typhoid fever caused by *S. typhi* appear in 8 to 14 days after eating contaminated food and last anywhere from 3 to 60 days. These include a fever of 104 °F, weakness, lethargy, abdominal pain, coughing, nose bleeding and enlarged organs. Typhoid fever is a serious illness that can result in death.

Giardiasis

Giardiasis is an illness caused by the parasite, *Giardia intestinalis* also known as *Giardia lamblia*. It is most common among children but may affect adults also. It causes atrophy and flattening of the villi in the small intestine resulting in less absorption of digested food in the intestinal tract. Symptoms of illness may start after two days of infection and include violent diarrhoea, excessive gas, greasy and floating stool, besides stomach/ abdominal cramps. The upset stomach is generally accompanied by nausea and vomiting. Lactose intolerance is commonly observed during the acute phase of the disease and can persist even after the eradication of *Giardia* from the digestive tract.

Giardia lives inside the intestine of infected humans or other animals. Individuals become infected through ingesting or coming in contact with contaminated food, soil, or water. The reservoir of *Giardia* parasite is contaminated items and surfaces that have been stained by the faeces of a diseased person or carrier as well as if they are involved in preparing and serving the food in houses or restaurants. When a person becomes infected by consuming contaminated food or water, some individuals will show signs and symptoms of the disease, while others will not and remain as carrier. The later continues transmitting the culprit parasite to healthy persons due to bad hygiene and sanitation practices for years together or may be life long as they don't approach physician in absence of any illness.

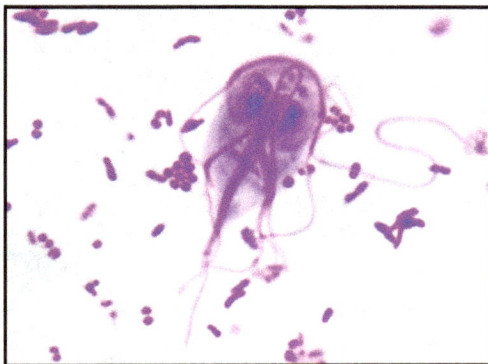

Fig. 2.2: Giardia

(Source: https://i.ytimg.com/vi/3uQRB0EvA34/maxresdefault.jpg)

Giardiasis is a mild diarrhoea but can also lead to dehydration because of the loss of fluids and electrolytes if individual has severe diarrhoea and vomiting, although it is rare. In most cases, symptoms last for 1-2 weeks. In some cases, symptoms will go away naturally due to washing off parasite from the intestine and may reappear a few day or weeks later as residual parasite in the intestine multiplies and again reach to the threshold level of illness. In cases where safe drinking water is not available, boiling of water for few minutes is one of the most effective method to make the water safe to drink and kill disease-causing agents.

Amoebiasis

Amoebiasis is commonly caused by *Entamoeba histolytica* which is an anaerobic parasitic protozoan. About 480 million people are infected with *E. histolytica* and this results in the death of 40,000–110,000 people every year across the globe. Amoebiasis is generally transmitted by the faecal-oral route, i.e., human faeces containing amoeba cysts get entry into gastrointestinal tract by oral route mostly through food and water. But it can also be transferred indirectly through contact with dirty hands or objects. Mostly, the infected people (about 90%) remain asymptomatic without any sign of illness (carriers) but transmit the parasite to others. Other mammals such as dogs and cats can also become infected transiently, but they do not contribute significantly to transmission of the disease. *E. histolytica* parasite remains in two forms during its life cycle; the active stage, *i.e.,* trophozoite (vegetative form) exists only in the host and in fresh loose faeces whereas dormant and hardy form, i.e., cyst (resting form) survives outside the host in water, in soil, and food especially under moist conditions. When cysts are swallowed they cause infection after conversion to the trophozoite stage in the digestive tract. Any non-encysted amoebae (trophozoites), die quickly after leaving the body and are rarely the source of new infections. Since amoebiasis is transmitted through contaminated food and water, it is often prevalent in regions of the world with limited modern sanitation systems.

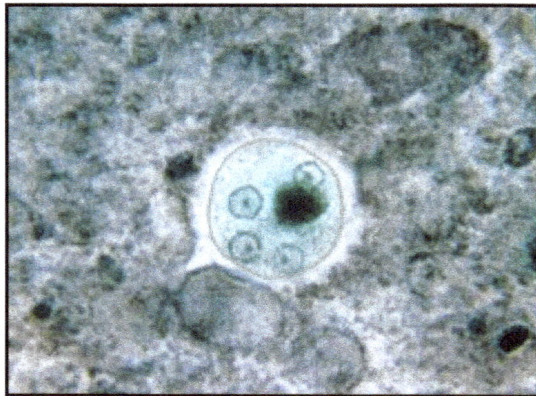

Fig. 2.3: *Entamoeba histolytica*

(Source: https://o.quizlet.com/3Zhz6laTm0lQ9utYuW16rQ_m.png)

E. *histolytica* infection can lead to amoebic dysentery and also the amoebic liver abscess. Although, symptoms of disease take from a few days to a few weeks to develop but commonly it is about 2-4 weeks. The symptoms include dysentery, bloody diarrhoea, weight loss, fatigue and abdominal pain. The affected people may develop anemia due to excessive loss of blood. Infections usually last for years and may stay even for the entire life. The amoeba can actually penetrate into the intestinal wall, causing lesions and intestinal symptoms, and it may reach the blood stream affecting the other organs usually the liver, but sometimes the lungs, brain, spleen, etc. A common outcome of this invasion of tissues is liver abscess that can be fatal, if untreated.

Cryptosporidiosis

Cryptosporidiosis is a parasitic disease caused by a protozoan, *Cryptosporidium*. It spreads through the faecal-oral route and Infection occurs through contaminated material such as soil, water, uncooked or cross-contaminated food contaminated with faeces of an infected individual or animal. It is especially prevalent amongst those who are in regular contact with bodies of fresh water including recreational water such as swimming pools. Other potential sources include insufficiently treated water supplies, contaminated food, or exposure to faeces. The high resistance of *Cryptosporidium* oocysts to disinfectants such as chlorine bleach enables them to survive for long periods and still remain infective.

The parasite causes an acute short-term, self limiting, diarrhoea by infecting the intestine. In immuno-compromised individuals, such as AIDS patients, the symptoms are particularly severe and often fatal. Symptoms appear within 2-10 days of infection, with an average of 7 days, and last for up to two weeks, or in some cases, up to one month. Diarrhoea is usually watery with mucus. There is often stomach pain or cramp with low grade fever. Other symptoms include nausea, vomiting, malabsorption and dehydration. Even after the symptoms of disease have subsided, the individual may harbour the parasite in the intestine and continue shedding the organism into the faeces for a number of days. Asymptomatic individuals (those with no symptoms) are not less infective than the patients and can similarly pass on the infection to other healthy individuals.

Hepatitis A

It is a liver disease caused by the hepatitis A virus (HAV). It can affect anyone irrespective of age and sex and can occur as isolated case or in widespread epidemics. Hepatitis A virus is found in the stool of persons infected with hepatitis A and mostly spreads from person to person. The virus is more easily spread in areas where there are poor sanitary conditions or where personal hygiene is poor. Persons with hepatitis A can spread the virus to others who live in the same household. Hepatitis A virus spreads when water sources are contaminated with infected human faeces. The virus can enter the water through various means, including sewage overflows or broken sewage systems and contaminated aquifers.

Three out of four adults who are infected by HAV will develop symptoms, usually over a period of several days. Symptoms generally appear 2 to 6 weeks after infection with the virus. The most common symptoms of hepatitis A include: Jaundice or yellowing of the skin and eyes, dark urine, fatigue, loss of appetite, nausea, vomiting, fever and stomach pain.

Hepatitis E

Hepatitis E is caused by infection with the hepatitis E virus. It spreads most often by contaminated drinking water and occurs mainly in developing countries where human waste gets entry into drinking water due to no or improper treatment.

The virus may start dividing in the gastrointestinal tract, but it grows mostly in the liver, after an incubation period of two to eight weeks. Typical signs and symptoms of hepatitis include jaundice, loss of appetite, an enlarged liver, abdominal pain, nausea and vomiting, and fever. Most often the illness is mild and disappears within a few weeks with no lasting effects. However, on rare occasions the acute illness damages and destroys large number of liver cells that impairs the liver function leading to death.

Polio

Polio (Poliomyelitis) is a communicable disease, which is categorized as a disease of civilization. It is caused by polio virus which is an RNA virus. It colonizes the gastrointestinal tract particularly the oropharynx and the intestine. The incubation time of the virus varies from 3-35 days before the appearance of first symptom. The polio occurs in humans only and especially affect young children below 5 years of age.

Polio is highly contagious and spreads through the faecal-oral and the oral-oral routes. Virus particles are excreted in the faeces for several weeks following initial infection. In areas with poor hygiene and sanitation conditions, the polio virus spread is fast and easy. In endemic areas, wild polio viruses can infect virtually the entire human population. It is seasonal in temperate climates, with peak transmission occurring in summer and autumn, however, in tropical climates it is transmitted uniformly throughout the year.

Symptoms of the disease include general weakness, fever, paralysis, severe breathing problem, and inability to use of one or both legs. The disease is usually fatal if the nerve cells in the brain are attacked (bulbar poliomyelitis), causing paralysis of essential muscles, such as those controlling swallowing, heart beat, and respiration.

According to the World Health Organization (WHO), one in 200 polio infections result in permanent paralysis. Though, the disease has been largely eradicated due to the development and widespread use of a polio vaccine among masses, the danger still prevails as even a single case of polio may transfer the infection to unimmunized healthy individuals especially in developing countries where sanitation is still of high concern.

Above mentioned diseases are some of the very common illnesses caused by human excreta that gets entry into the human gut through water and food, unhygienic practices, dust or carriers. The list is just indicative and is not exhaustive as there are hundreds of pathogens and parasites which infect the people through faecal-oral route and may be difficult to include in this text. In the nutshell, there is no alternative to sanitation which otherwise leads to big burden on the society in terms of health, sufferings and economics. This is also evident when the comparison is made between developed and developing countries.

3

Efforts Made by Indian Railways for the Disposal of Human Waste

Indian Railways(IR) has its network spread throughout the country and is the most reliable mode of transport for more than one billion people. Over the years, it has made appreciable development for spreading the rail network, number of trains and modernization of the coaches. However, visible impact is lacking with respect to hygiene and sanitation that not only tarnishes the image of this government-owned carrier but has also become a threat to the health of its passengers as well as the residents of colonies located near to the stations and railway tracks. Proper human waste disposal has always been a major concern for the railways.

Presently, human waste generated in the coaches of Indian Railways (IR) is discharged directly on to the railway track. Indian Railways has taken measures to stop this open discharge by experimenting with Controlled Discharge Toilet System (CDTS) especially at Railway stations and populated areas of cities. The focus has been to maintain the cleanliness of the railway stations rather than railway tracks irrespective of their closeness with the residential areas. Two reasons can be assigned for this thought; firstly, railway stations get the maximum faecal matter discharge as people prefer using the toilets when the train is stationary, and secondly, stopping the discharge at stations make a clear and visible impact of the efforts made. Other efforts made by railways have gone for onsite treatment of the waste generated from each toilet so that effluent is free from objectionable odour, organic matter and pathogens and it can be discharged on the track, irrespective of stations, cleaning sheds or cities.

Controlled Discharge Toilet System (CDTS)

The CDTS is based on German technology and the basic purpose of its introduction in IR was to eliminate the practice of discharging toilet waste on to railway station area and in the populated area of city where the train does not attain a predefined speed. A controlled discharge system as opposed to conventional systems aims at:

- Clean, odourless, hygienic and aesthetically pleasing toilets
- Non dumping of human waste at stations and surrounding area
- No spilling of waste on rails and coach parts like under gears

These toilet systems are designed to operate on the principle of high -flush through which the evacuation of toilet bowl is carried out by means of water pressure. It operates with a pressurized water bowl wash that covers 100% of the toilet bowl area.

Salient features of control discharge toilet system are as follows:

- It is easily programmable
- Less requirement of air, water and electricity
- Fully PLC Controlled
- Easy to clean

In the CDTS, waste is removed from the toilet bowl and transferred to a retention tank located under the toilet bowl with a minimal amount of water. Water consumption is only 2.5 litres per flush cycle for the Indian style toilet and 1.5 litres for the European style toilet. The pressuriser provided in CDTS system, delivers pressurized water to flush the waste. The capacity of retention tank is 45 litres. The retention tank which stores effluent has two openings- one at the top gets the waste from toilet and another at bottom to discharge the waste on railway track. These two openings get activated by double acting pneumatic cylinders fed by Feed pipe of air brake system. Upper opening gets opened every time the user operates the flush button, whereas lower one opens only at predetermined speed of the train and after predetermined number of flush cycles. The solenoid valves control the entry of air in pneumatic cylinders attached to openings/slides.

Fig. 3.1: LHB coach fitted with controlled discharge toilet unit

(Source: https://i.ytimg.com/vi/yER32dk28N0/maxresdefault.jpg)

By pressing the flush button, a passenger activates a flush cycle, where water flows into the toilet bowl and the slide valve connecting the toilet bowl to the waste retention tank gets open. At the end of each flush cycle the wash water stops and the toilet is sealed off from the retention tank by the slide/flapper valve. This valve acts as an effective stench trap, preventing odour from entering the toilet room. The waste is stored in the retention tank until two conditions are met; 1. a predetermined number of flush counts have been recorded, and 2. the train has reached a minimum speed of above 30 kmph. When these two conditions are met, the retention tank's discharge valve opens and the waste is expelled, away from railway stations and heavily populated areas. The retention tank's discharge valve remains open only for one minute, which is long enough to completely drain the tank.

Thousands of CDTS units were fitted by IR in different long and short journey trains but technology could not sustain for long and now almost is given up by Railway authority. Both technical and social (passenger's awareness and habits) reasons have been responsible for such failures. Some of the important reasons that forced Indian Railway to give up this system are:

(a) Passengers are used to throw the garbage in the toilet pan since toilets are not provided with dust bins and mostly pans of trains also work as open garbage disposer. Large garbage items dumped into the toilet, especially plastic bottles and napkins, block the toilets very frequently.

(b) Excess accumulation of the waste in the retention tanks leads to overflowing. The tanks were designed for 45 litres and they would fill up with regular water wasted even if the toilet is not used for relieving.

(c) As there is no mechanism and infrastructure in place for cleaning of these toilets, they are cleaned in the pit lines. Flooding of pit lines with night soil during cleaning has been a major point of concern. This results in night soil accumulation in the pit lines and creating unhygienic environment for the staff.

d) Sometimes programmed opening and closing systems do not work due to technical problems that leads to overflowing of water as well as waste.

Further, there have been issues with the technology itself like human waste is still discharged on the rail track barring only a few stations where train stops. Sometime train attains the minimum speed of 30 kilometer per hour on the longer platforms and if not from all coaches but at least from those which are positioned on the back, the waste get discharged on the railway station. Some trains, not having stoppage at certain stations, continue discharging the waste at those stations. Moreover, big cities like Delhi has huge number of habitants spread over 30-40 Km of distance and it has closely located, in continuation, other cities like Ghaziabad, Gurgaon, Noida etc. Hence, controlling the human waste discharge at one or two Railway stations is hardly of any use.

As per the Integrated Railway Modernization Plan (IRMP) 2005-10, CDTS were to be installed in 5000 coaches by March 2010. However, due to problems in flushing system and dropping of discharge more or less at the same locations, implementation of CDTS was restricted to only few trains like Rajdhani, Shatabdi and Duranto. Presently, no CDTS installations are taking place in any of the trains.

CDTS, no doubt, may be an excellent technology for developed countries having less dense population, less crowded trains, different living culture and better literacy rate, but could not prove its worthiness in Indian scenario because of above cited reasons and maintenance-related issues. Further, the cost of CDTS is too high. The expenditure required for four toilets in a coach is 7-8 lakhs and this along with maintenance cost (frequent maintenance needed) has also restricted the application of technology in Indian Railways.

Zero Discharge Toilets System (ZDTS)

This toilet system has been developed jointly by Research Design and Standards Organization (RDSO) and Indian Institute of Technology (IIT), Kanpur. ZDTS work on the principle of solid-liquid separation with solid part being stored-evacuated-transferred and dumped in to pits for composting and the liquid portion filtered-treated-recycled for flushing purposes.

The key component involved is a unit called a "Solid-Liquid Separator" (SLS). This unit is connected to the P Trap of any regular commode (both Indian/ Western style) at the discharge end. It is a cylindrical structure and has a dimension of 2 ft diameter and 2 ft height. So, depending on the configuration of the commode, this unit has to be fitted either under or behind the toilet. The SLS cylinder has spiral perforated grooves coming down from top to bottom on inside. When the toilets are flushed, the vortex movement of water cleans the pan surface and pushes the solid waste downwards into a tank at the centre. The centrifugal force act outwards from a centre of rotation and presses water to come out. The geometric design of the surface guides it through a circular path downwards towards the separator. At the separator, the water is guided into pipes in the sides that take it to another tank. Thus the solid and liquid is separated right at the source. A very small quantity of the liquid goes along with the solids, but that aids it in moving along the solids pipe line to a collection tank as slurry. It is claimed in journal 'Down to Earth' in its January 2008 issue that the separation of water from the solid is close to 90%. The liquid is collected, filtered through a simple membrane filter and re-circulated for flushing. The pipes are fitted with micro-filters made of high quality poly vinyl chloride which clean the liquid. There is a microbicidal liquid added on daily basis that completely takes care of the odour and aesthetics of the flush water.

Inventors have proposed that the tank containing solid waste will have exit points that could be connected to pipes. The solid waste can be sucked out and transferred into containers or trolleys. In the railways, once the train moves into the yard, pipes could be attached to these exit points and the tanks emptied. The solid waste can be converted into useful compost by mixing with old compost followed by treatment with worms. The trolleys will dump the waste into a unit identical to a concrete mixer, which will already have some compost in it. Once it is mixed properly, it will be dumped outside and allowed to decompose for sometime on its own. This pre-compost is further added to the fresh load of solid waste and the process repeated. After several cycles, the concentrated compost is ready for vermicompost. The solids will be put up through a second stage composting process, where vermi-culture is done using earthworms (*Eisinia foetida*).

The final product is claimed to be completely pathogen-free and odourless, and a soil conditioner. The excess of liquid post flushing usage may be sun dried to produce water-soluble crystals rich in urea & potash. The collection of solid waste and liquid overflow at stations by suction through pipelines and their conversion into useful products like compost and water soluble crystals rich in urea and potash remained only the claims and could never be demonstrated either at small or large scale.

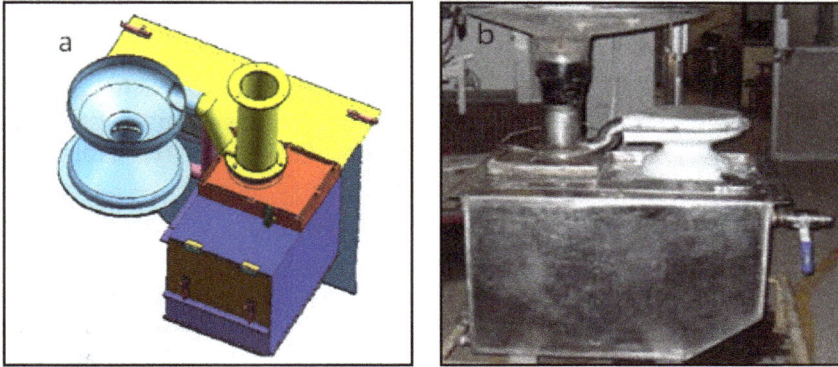

Fig. 3.2: Zero discharge toilet system model (a) and prototype (b)

(Source: http://www.iitk.ac.in/olddord/TMRS_Status%20report_DORD.pdf)

A professor working on the project at IIT, Kanpur explained the benefits of such toilets, "It will avoid filthy conditions on the track; prevent corrosion of the underside of the toilet portion of the coach and rails leading to substantial savings; and maintain hygienic conditions. The estimated cost of installing ZDTS is approximately Rs. 10 lakh per coach, and the annual operating and management cost is Rs. 35,000 per toilet. "He further added that the Railways would have to spend a substantial amount to ensure that proper infrastructure is in place (The Sunday Guardian, March 03, 2016).

Trials have been conducted on one coach on a train running between Chennai and Lucknow, and Chennai and Jammu Tawi. RDSO planned to implement the ZDTS on 14 coaches of a train from the southern region but it could not be done. Despite RDSO reporting the performance of this model as satisfactory, the project was abandoned in September 2009, as part of closure of TMRS projects and prototype toilets were removed.

There have been many reasons for failure of ZDTS in Indian Railways, some of them are described below:

- Crux of the technology is separation of water from human solid waste and separation is claimed to be 90-99%, however in reality, it is much lower. More water in solid storage tank may lead to overflow in case if number of passengers are more than the expected (that usually happens in Indian trains), if train takes longer journey period than scheduled due to some delay or if place of cleaning the tank is far off.

- A few percent of waste solids (especially soluble solids) go with the water which is going to be reused for flushing. Masking the colour and off odour don't make water free from pathogens and aesthetically acceptable.

- The quantity of water used per person for ablution and washing and the amount of urine generated per defecation will not be less than 2L. Where this excess water will go? Further, some passengers will continue using toilet for urination (4-5 times a day = 1L). This additional liquid (either generated or used) has to be discarded on rail track as against the claim of zero discharge.

- The discharged liquid (overflow) is expected to be of worst quality as it will have soluble faecal matter with large number of pathogens and will stink badly besides risking the corrosion of tracks, thus defeating the basic purpose of taking all pains.

- There is requirement of a device and energy for pumping the liquid to overhead tank.

- The infrastructure needs to be created at stations to evacuate the toilets. It requires additional funds, space and manpower. There can be failures of evacuation device (technical, chocking due to non biodegradable material added to toilet) and spilling of waste.

- Arrangements are needed for composting/ processing of solid waste and excess liquid waste. If it is at stations, processing will lead to foul smell in the surrounding and in case of far off place, transportation in close containers is a must which will involve additional money.

- The cost of ZDTS is too high, i.e., for each coach Rs 10 lakhs initial and Rs 1.40 lakhs annually for maintenance. When added with the cost of creating infrastructure for evacuation of accumulated solid waste and its subsequent treatment, certainly the technology becomes unacceptable for a country like India.

Thus, ZDTS cannot be considered better than CDTS from any angle. It is neither cost competitive nor has been subjected to pilot field trials as done for CDTS. The trials conducted with one or two toilets only for segregation of solid in absence of further treatment (only speculation) does not mean anything.

Indian Railways has also worked with other technologies which aim at treating the waste on site, i.e., as and when waste is generated. The human waste is biodegraded and converted into gases and water by employing useful aerobic/ anaerobic bacteria. In addition to removal of organic waste, the biodegradation process inactivates most of the pathogens also. And the remaining ones are subjected to further treatment with chlorine before effluent gets discharged on the track. The best part of such technologies is that being a continuous process, there is no need of imposing restrictions on passengers for using the toilets at railway stations (as it is presently written in every toilet: Please Don't Use Toilet in Stationary Train) since treated effluent that get discharged is expected to be odourless and safe. The two such technologies attempted by Indian Railways are: Microphor bio-toilets and biodigesters developed by Defence Research & Development Organisation.

Microphor Bio-toilets

These have been developed by California-based Microphor, a subsidiary of Wabtec Corp. The toilets work on "dry-bed" principle where solid waste is consumed by bacteria while liquids are discharged after chlorination. The treatment tank is composed of three chambers; top chamber contains vertical fibre columns mounted on floor for biological treatment, middle chamber is meant for collection of biologically degraded waste as liquid, and a bottom chamber where chlorinated liquid gets sufficient time for disinfection. The secondary retention chamber has baffles to provide longer treatment path to the liquid for adequate chlorination. The chlorinator is an attachment to the main tank and has provision of keeping the slow release chlorine tablet (approx. 7.5 cm diameter x 2.5 cm height) through which biologically treated liquid passes.

When waste water from the toilet enters the top chamber, it comes in contact with bacteria (aerobic) and enzyme mixture. The bacterial- enzyme mixture is patented, imported, and powdered formulation that is added to the top chamber in defined quantity at fixed interval. Waste and bacterial mixture passes through vertical columns, horizontally, and liquefied waste travels down to the middle chamber through pores/ slits in columns. The liquid, then, passes through chlorinator and gets some chlorine dissolved. It is the active chlorine in free form that is actually responsible for killing the pathogens. Secondary retention chamber provides sufficient time for chlorine to act on pathogens and kill them. The waste liquid in the treatment tank moves on gravity flow basis and is finally discharged on the railway track.

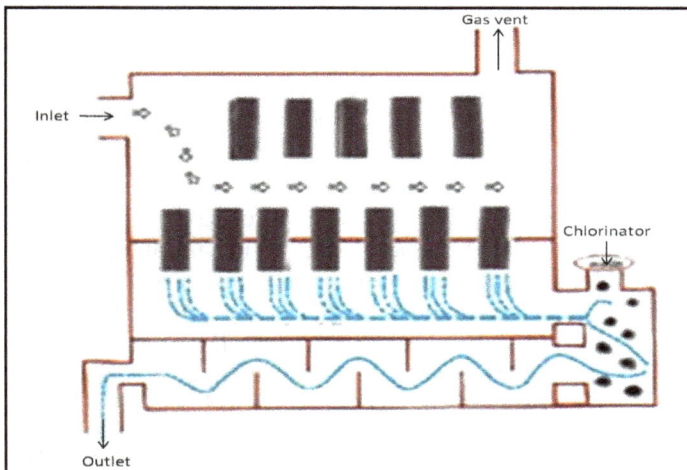

Fig. 3.3: Material flow in Microphore biotoilet

(Source: http://www.downtoearth.org.in/coverage/indian-railways-experimentation-with-ecofriendly-toilets-4386)

The Indian Railways, during the first initiative to stop the open discharge of toilet's waste on tracks, evaluated these bio-toilets, way back in 1993. Southern Railway introduced eight coaches fitted with Microphor bio-toilets into service. The initial feedback from passengers was encouraging, however, shortly, complaints started pouring about foul smell from tanks, faecal matter visible from the top, cockroaches and flies infestation, clogging and the leakage from tanks. Due to repeated complaints, the bio-toilets were discontinued in1995 and removed from the coaches. Integral Coach Factory (ICF), Chennai took active interest in the technology and suitable modifications/ suggestions were made before retrial, like-

- Removal of cross wire in the chute meant to prevent cloth and solid material which otherwise was getting entangled.

- Slanting of chute so that faecal matter is not visible from the top.

- Providing exhaust fans in toilets and their continuous operation for elimination of foul smell

- A separate waste chute for cups, bottles, plastics etc.

- Prohibiting the usage of phenol, chemicals and strong detergents in toilet

Following this, 10 AC coaches were fitted with Biological Toilets and put again into service during December 1995 to May 1996. These worked on 'wet bed' principle, requiring the bacterial agent, cleaning agent and a stain remover as consumables. Again, passenger's feedback collected initially was encouraging but afterwards complaints started pouring as enumerated below-

- Bad odour from tank

- Leakage from flange joints

- Insects and flies breeding in the tanks

- Tanks getting filled by non-biodegradable waste thrown into the toilets. Removal of these on pit lines led to unbearable stink in the pits and objection from the staff.

Susequently, more corrective actions were taken but bio-toilet could not prove its worthiness.

Southern Railway workshop helplessly removed the systems from all coaches in 1998. Railways continued exploring other viable technologies for waste disposal for almost a decade but seeing no other option, in January 2007, Railway Board again placed a development order for the similar aerobic bio-toilets on M/s Aikon Technology Limited (Indian Representative of Microphor, USA) for design, manufacture, supply, installation, commissioning, maintenance and operation of 80 bio-toilets. These prototype bio-toilets were procured and installed on Prayagraj Express and Rewa Express. During the trial period, the supplier carried out a number of modifications. There were, however, several complaints of bio-toilet getting chocked, excreta over flowing and directly falling on the track, besides, the effluents not complying with the stipulated test parameters.

These aerobic bio-toilets were in news for a period of more than two decades and during this time, they were subjected to many modifications by industry/ IR and repeatedly evaluated by IR. Now, evidently it became very clear that technology was not at all suitable for Indian passenger trains. Passengers' awareness and prevalent practices might have been responsible for the failures, to a certain extent, but major reason could be assigned to the flaw in basic technology like non mixing of bacteria/ enzyme with the excreta and insufficient oxygen for the optimum activity of the aerobic microbes beside chocking of pores in the matrix (vertical fibre columns) by undegraded waste in top chamber.

Further, the technology was cost and maintenance intensive. For every coach, the railways was to shell out Rs 8 lakhs as equipment cost and Rs. 1.5-2.0 lakhs as operations cost per year. Bacterial powder that is to be added frequently is patented, not available indigenously and need to be imported every time. Hence, this technology did not stand any chance of adoption by IR and, thus, railway authority went ahead with other alternatives.

DRDO Biodigester Technology

Defence Research & Development Establishment (DRDE), Gwalior is one of the premier laboratories of Defence Research & Development Organisation (DRDO) under Ministry of Defence, India. It had developed biodigesters for high altitude-low temperature (as low as -40 °C) and plain areas for army and civilians. And the focus was on stationary mode of application for human waste disposal. As Indian Railways was struggling, since 1993, to solve the problem of open excreta discharge on tracks, a meeting was arranged at Sulabh International complex Delhi during 2004 involving different private industries and government institutions working in the area of sanitation. In the background of Indian Railway's experience for the last one decade on the subject, detailed deliberations were held to work out a suitable technology solution of waste disposal in passenger coaches. DRDE was entrusted with the responsibility to develop the biodigester technology for passenger coaches as it had vast experience in this field.

Although, DRDE was having experience of developing biodigesters for much tougher geo-climatic conditions of extreme low temperature like Siachen and Leh, but this time the challenge was developing the technology for mobile applications. Further, the number of users was variable, space was limited and ambient temperature was to change from one extreme to another with varying time interval during train journey. The technology challenges are enumerated below:

- Number of passengers using a toilet varies with the type of coach (AC-I, II and III; sleeper and general) and the type of toilet seat (Indian/ Western) from a few number to 30 or even more.

- Train journey varies from few hours to a few days and, hence, the toilet usage.

- Ambient temperature is different in different areas of the country. Some trains operate continuously at almost similar temperature throughout the year (low or high) whereas others face seasonal and locational variation.

- Some trains go to areas having ambient temperature close to °C in north (winters) and 45 °C in south during same journey thus getting wide temperature variation on the same day.

- The space available beneath the toilet is limited that restricts the volume of biodigester for defined number of users.

- Use of routine toilet cleaning detergents and antiseptics like phenyl by Indian Railways which is supposed to adversely affect the activity of useful bacterial population in the biodigester.

- Use of excess and uncontrolled quantity of water by operation and maintenance staff during cleaning of the toilet that might wash out the useful bacterial population from the treatment tank.

DRDE undertook the laboratory investigations considering above mentioned concerns like working of anaerobic bacterial consortium under expected temperature variables, in presence of detergents and phenyl and reduced treatment time (hydraulic retention time, HRT). Use of bacterial immobilization matrix was also explored. Various designs of biodigester were considered and evaluated. Finally, prototype biodigester was subjected to rigorous testing in the laboratory before actual trial in passenger coach. Getting convinced with the performance of biodigester, Indian Railways decided to use the technology for eradication of sanitation problem in all of its trains.

4

Development and Induction of Biodigester in Indian Railways

DRDE has been working since 1989 for the development of a technology for eco-friendly disposal of human waste. Focus of the scientists had been high altitude and low temperature areas of northern Himalayas including Siachen Glacier with army as primary user. Anaerobic Biodegradation Technology, nicknamed as Biodigester, was developed and successfully demonstrated in different areas having varied temperature ranges. Glaciers' ambient temperature remains sub-zero (approx.-40 °C) throughout the year whereas in Leh (Ladakh), it varies from -28 °C to+28 °C depending upon the season.

Various reasons can be ascribed to for opting the anaerobic process as against aerobic method for human waste biodegradation but the most important being the closed system which helps in retaining the temperature inside the fermentation tank by preventing heat losses to the outside sub-zero temperature environment. It is mandatory for the substrate (waste) in the tank to be in molten state for its biodegradation as bacteria will not eat away the waste in solid state (frozen) and closed systems are the only viable system. Exothermic biochemical reactions of bacteria during fermentation of heterogeneous waste generate substantial calories inside the tank. The remote cold areas lack conventional energy sources which is essentially required for agitation and aeration of the waste for aerobic fermentation thus tilting the choice further in favour of anaerobic digestion.

A strategy was made for biodegradation of human waste at such extreme low temperature areas by designing suitable digestion tanks with appropriate insulation and heating arrangements along with consortium of bacteria for such extreme low temperatures. Biodigester were installed and made functional at different high altitude locations. Subsequently, biodigesters were developed for civilians living in varying geo-climatic conditions of plains as well. The size and material of construction of biodigesters varied depending upon number of users, climate, connectivity of the area, availability of water and socio-economic status of people. All these developments, however, were undertaken for the stationary applications in rural and urban sectors.

Fig. 4.1: Temperature controlled biodigester in snow bound areas

Development of biodigester for mobile application (Indian Railways) required a different strategy and developmental approach. The efforts of DRDO to achieve the goal, over the years, for this purpose can be classified into different headings for ease of explanation and understanding. These are - Anaerobic Microbial Inoculum (AMI), immobilization matrix, biodigester designing, and laboratory as well as field trials of prototypes and final equipment, respectively.

Anaerobic Microbial Inoculum (AMI)

In its endeavour to develop the technology for bioremediation of human excreta at low temperatures, DRDE undertook the experimental work during 1990-1995 to develop the microbial inoculum, which is the most critical component. Developing the anaerobic microbial consortium for low temperature applications, however, was very tricky and tedious.

In nature, the saprophytic bacteria can broadly be divided into three groups based on temperature range of their growth and multiplication; 1. Thermophiles (optimal growth above 45 °C), 2. Mesophiles (20-45 °C), and 3. Psychrophiles (optimum growth below 20 °C). Some of the bacteria prefer mesophilic temperature but still can grow upto freezing temperature, they are called Psychrotrophs or cold tolerant. While designing the microbial consortium, it was possible to use the bacteria belonging to any of the temperature groups, however, each group has certain advantages/ drawbacks for their working at low temperature. Selecting and developing inoculum from thermophilic/ mesophilic group of bacteria would have been advantageous since these bacteria have higher rate of metabolism leading to fast biodegradation but maintaining the required temperature in the biodigester by external energy source is costly and tedious task. Psychrophilic bacteria can perform the task at much lower temperature thus needing little energy for the maintenance of temperature. But these bacteria are comparatively slower in action, thus, needing biodigesters of much higher volume hence require high cost and infrastructure for installation. The major drawback of choosing such bacteria is their inactivation/ death above 20 °C which is more likely to happen at places having wide temperature variations during summers and winters even at places like Leh besides limiting their application in plain areas with perennial high temperatures (above 20 °C). Hence, the best thought option was to develop the bacterial inoculum from psychrotrophs or cold tolerant microbes which can work in wider temperature range.

The anaerobic microbial inoculum (consortium) has been developed by enrichment of bacteria, from biogas plants working in hilly and low temperature regions. The initial inoculum was taken from different biogas plants operational with cow dung, kitchen waste and other agro wastes in low temperature areas of Himachal Pradesh. The most efficient inoculum was adapted with the desired substrate, i.e., night soil or human waste. The 20 °C grown inoculum in presence of human waste was subsequently adapted, gradually, to low temperature by decreasing the temperature at the rate of 2 °C every quarterly till 10 °C, and at the rate of 1 °C till 5 °C. The adapted culture had been further improved for its performance by bio-augmentation with independently isolated bacteria, belonging to proteolytic, acetogenic and methanogenic groups, from various places. This consortium was evaluated for its performance in the laboratory and found suitable for Railways. The bacterial consortium, so developed, is being multiplied and maintained continuously at DRDE.

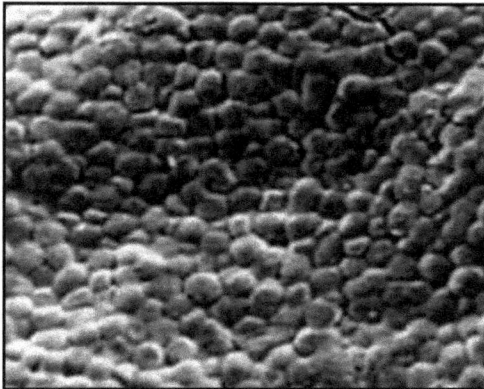

Fig. 4.2: Electron micrograph of bacteria from anaerobic microbial inoculum

Toilet cleaning agents and antiseptics are well known for their antibacterial properties and, thus, their use consequently affect the biodegradation process. Maintenance staff of Indian Railways routinely use detergents and cleaning agents including phenyl in the toilets of coaches following defined schedules. The effect of these chemicals, on biodegradation efficacy of bacteria, has been evaluated in the laboratory. None of the commercially available chemicals indicated any adverse effect on the fermentation of excreta even at double the concentration of their specified use.

The temperature fluctuation is known to adversely affect the biodegradation (bio-methanation) process and it is bound to occur at different magnitude and frequency in trains as they travel long distances with different climatic conditions. Under such conditions, the performance of bacteria developed at DRDE was evaluated and it was observed that temperature fluctuation certainly affects the metabolic process, but to a tolerable level and for short duration only. This effect was taken care by making suitable provisions in biodigester design and by incorporation of microbial immobilization matrix in the biodigester.

Bacterial inoculum addition into the biodigester is a pre-requisite before its use by the people. Different volumes of inoculum were tested for initiating the biodegradation process ranging from 10-50%. The addition of inoculum quantity at the rate of 30% (v/v) to the total volume of the tank was found to be sufficient for the optimum performance.

Hydraulic retention time (HRT) is a measure to know the rate of biodegradation of any organic material and, accordingly it dictates volume requirement of biodigester for defined number of users. Low HRT needs lower volume of the tank because of fast action of the bacteria and higher HRT requires higher volume of the reactor due to slow rate of waste decomposition in the tank. Because of space constraints, low volume biodigesters (low HRT) could only be accommodated beneath the toilets of coaches, hence the biodegradation efficiency of bacteria was tested at 1-5 days HRT. It worked well at 2-5 days for continuous and long term use and at two days for the short term.

Immobilization Matrix

Biodegradation of the waste is expected to be proportional to the number of bacteria involved in the biodegradation process. Although, most of the bacteria involved in waste degradation, remain in free form (floating/ swimming) in the biodigester, but a certain population get attached to waste particle till particle gets digested and are, generally, not washed away with the effluent. Free bacteria are maintained at a constant number in the biodigester that means the additional bacteria produced in the tank get washed away with the effluent. Thus, biodegrading bacteria in immobilized form in the biodigester have added advantage as they are retained in the tank as additional bacteria contributing for faster bio-degradation. Further, such bacteria showed better tolerance for adverse conditions like temperature fluctuations, cleaning agents and toxic chemicals generated (like propionate) in the biodegradation process.

The anaerobic biodegradation (bio-methanation) process used in the biodigester is a four step sequential process where end-product of first step is a substrate for next step till the waste gets converted to water and gases. The immobilized bacteria of diverse group reside in close vicinity, therefore, they carry out multi-step process of biodegradation more efficiently. Thus, immobilization of bacteria was considered very essential for developing the technology of mobile systems.

The large number of immobilization matrices were screened keeping in view the criteria of better bacterial attachment, that included durability, large surface area and cost. Poly vinyl chloride (PVC) based matrix having high surface area was selected and tested in the laboratory for performance and its application in the biodigester.

Fig. 4.3: PVC matrix for immobilization of bacteria present in anaerobic microbial consortium

Designing of Biodigester

Based on the performance of anaerobic microbial inoculum (microbial consortium) in terms of hydraulic retention time (HRT), the volume of biodigester for the Railways was calculated. Although, the suitable HRT was worked out to be 2-3 days, but for practical purposes the tank volume was calculated with 5 days HRT considering the over usage of toilets at certain situations and higher volume biodigester was decided. This also takes care of slowing down of bacterial activity at low temperature or in presence of excess cleaning agents. The increased volume accounted for the safety margin of approximately 100%. It is worth emphasising that excess input of excreta into the biodigester tank than the calculated amount, leads to souring and failure of biodegradation process which cannot revive until tank is cleaned and refilled with fresh bacterial inoculum.

The first biodigester was designed to cater the waste output of two adjoining toilets in a coach. It was fabricated using mild steel of 4 mm thickness. The Biodigester tank had mainly two compartments (chambers) - one for biological treatment with bacteria and another for chemical disinfection with chlorine. The first chamber had five partitions for the waste to travel following longer path for degradation in presence of bacteria till it enters into another chamber for disinfection before final discharge on to the ground. All partition walls and internal surfaces coming in contact with bacteria were fitted with PVC immobilization matrix having large surface area. The biodigester in the centre had groove on the top of the structure to allow electric wires and pipes passing from one coach to another. Thus fermenting liquid waste traversed from first half of the tank to the another half from the lower restricted passage. There were two gas outlets from opposite ends of biodegradation chamber which finally got converged in to a single outlet. The designs of these biodigester tanks have been patented. (Patent no. CN 200780032599 (People's Republic of China), AU200727070 (Australia), 2009/00239 (Republic of South Africa), PCT/IN2011/000318, PCT/IN07/00278)

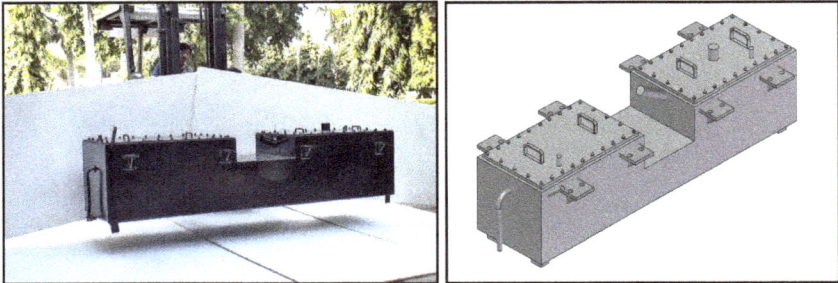

Fig. 4.4: Prototype of double toilet biodigester

Approximately, one-sixth of the total volume of biodigester was devoted to disinfection tank, provided at one end. Chlorinator was fixed inside this tank and was comprised of a moving disc having a slot for chlorine ball. The disc attached to a float valve rotates based on the liquid level in the tank. One chlorine ball drops into the tank every time when liquid drops to a certain level. The quantity of chlorine in a ball is enough to disinfect the liquid in the tank. The treated effluent gets discharged out through a siphon arrangement when the disinfection tank is full with the effluent. This arrangement provided the optimum disinfection to the biologically treated effluent in batch mode without involving any electrical gadget and any manual intervention.

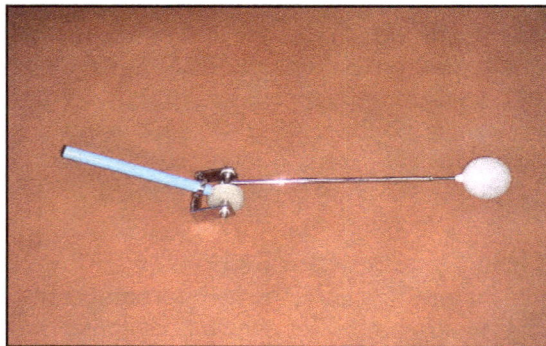

Fig. 4.5: Chlorine ball dispensing assembly

Field Trials of the Prototype under Stationary Condition

Above described biodigester was designed and fabricated for the purpose of field trials in a stationary mode rather than actual testing in mobile coaches and, thus, did not have required fixtures for fitment to the coach. However, functioning of siphon system and chlorination was tested in a moving truck. Two elevated toilets, replicating the scenario of two opposing toilets in a coach were constructed at DRDE residential area. The biodigester was placed below the toilets and connected at inlets for waste entry into the tank. The outlet of the effluent was connected to an existing drain available at the place.

Fig. 4.6: Double toilet biodigester installed under temporary toilet structure for testing under stationary condition; (A): View of biodigester under the toilet shelter, (B): View of toilet installed with biodigester

Fig. 4.7: Testing of siphon system of biodigester in a moving truck

The biodigester was charged with bacterial inoculum (30% v/v) and toilets were subjected to use by approximately 50 persons. Performance of the biodigester was evaluated at regular intervals by measuring biogas production, BOD, COD, pH, volatile solids, total solids, dissolved solids, volatile fatty acids, coliforms and off odour, if any. Trials were continued for almost 12 months to expose the biodigester under different weather conditions (temperature ranged between 3-47 °C). After the trials, biodigester was disconnected and opened for the physical observation. It was noticed that inside surface of the tank got rusted severely and PVC matrix pasted on inside surfaces peeled off at certain points. These matters were considered for design and fabrication of final version of biodigester for trials in railways.

Field Trials in Railways

Based on observations made during the laboratory trials of the prototype (mild steel biodigester), biodigesters were fabricated using stainless steel as prototype made of mild steel rusted badly in the short span of testing.

Two types of biodigesters were made; one for the single toilet (i.e, biodigester tank to be accommodated below each toilet) and the other for holding inputs from two opposing toilets. These toilets were fixed at either ends of the passenger coach.

The double toilet biodigester was of similar design as of prototype except that chlorination chamber was eliminated and in place of this an electronic chlorine dispenser was used.

Single toilet biodigester also had six chambers wherein the waste was to flow in zigzag manner, both vertically and horizontally, to provide the longer path for accomplishing efficient biodegradation. The immobilization matrix was pasted on all internally exposed surfaces and additional support to hold the immobilization matrix was provided by fixing a stainless steel wire mesh (1 sq inch) by nuts and bolts.

An electronic chlorinator, meant for dispensing hypochlorite solution at calculated dose and interval, was attached above the biodigester (on the end wall of the passenger coach). The liquid chlorine dispenser released chlorine into the last chamber of the biodigester. Arrangements were made to stop the back flow of chlorinated effluent into previous chambers to avoid the inactivation of bacteria by chlorine.

Fig. 4.8: Single toilet biodigester fitted beneath each toilet under the coach

Fig. 4.9: Double toilet biodigester connected with both toilets

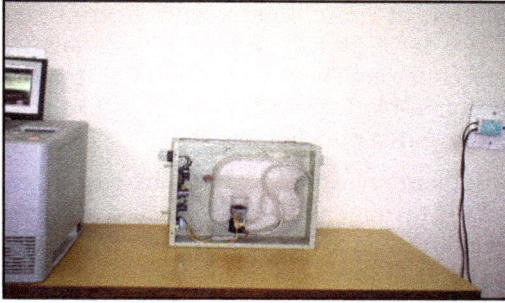

Fig. 4.10: Chlorination assembly for dispensing liquid chlorine in the biodigester

Both types of biodigesters were fixed under the coach by mounting brackets either sides along the length of the tank. Trials were initiated in Barauni Mail in the year 2005 after charging them with the microbial inoculum.

The Barauni Mail was operational between Gwalior (M.P.) and Barauni (Bihar) covering a distance of 1057 km. The scheduled departure of the train from Gwalior was at 1130 hours and the same train started its return journey from Barauni at 1845 hours next day, taking approximately 56 hours to complete the journey both ways. The performance of biodigesters was observed for a period of 20 months and the data were recorded for biodegradation parameters, effluent quality as well as physical observations. In addition, passengers feedback was also taken by providing them a questionnaire.

During some of the trips of Barauni Mail, toilet pan pipe was found blocked due to the presence of bottles, sanitary napkins, gutka pouches, etc. These chockings were cleared manually on need basis. Efforts were made to overcome the problem of chocking by exercising different options like providing cross wire, mesh and hooks in the toilet chute. The problem of chocking was overcome to a greater extent by cross wire and hooks in the toilet chute.

Fig. 4.11: Biodigester filled with non-biodegradable material as revealed after opening

After the completion of trials, biodigesters were opened and physically inspected. Unexpectedly, large chunk of foreign materials were observed in the first chamber of the biodigester. The foreign material found in the first chamber of biodigester included non-biodegradable materials like plastic bottles, sanitary napkins, polythene bags, gutka pouches, etc.

Analysis of the problem revealed that in trains passengers used to throw the garbage in the toilet chute through which garbage is disposed off onto the railway tracks as dustbins are not provided in the toilets. In other words, toilet chute also provides a place for disposal of waste in the toilets either generated inside the toilet or outside. This habit of passengers continued even after fixing of biodigester beneath the toilet as mostly passengers were not knowing that such a device has been beneath the toilet or they were not knowing that such practice will impede the functioning of the device. Even if some of them were aware about it, they did not have alternatives as no dustbins were provided in the specific toilets. Barring this, the performance of biodigesters was found satisfactory and encouraging to the scientists of DRDE as well as administration of Indian Railways. Therefore, trials were extended on pilot scale employing 36 biodigesters on trains like Magadh Express and Brahmputra Express.

Comparison of DRDO Biodigester Technology was made in parallel with other technologies, tried in different trains by Indian Railways with respect to the cost, efficiency and maintenance besides passengers feedback. It was found that the Biodigester Technology outscored the other Indian and foreign technologies tested by Indian Railways.

Convinced with the suitability of Biodigester Technology for addressing the problem of sanitation in trains, Indian Railways decided to implement the DRDO Biodigester Technology in its passenger coaches. Authorities of Indian Railways further took the decision to discontinue the trials of human waste disposal involving other technologies due to inherent problems like maintenance, cost (capital investment and recurring), requirement of additional infrastructure, etc. A series of meetings was held between IR and DRDO officials and it was decided to work jointly to implement the Biodigester Technology in all the passenger coaches of the trains and overcome the problem of chocking by design improvement in the interface of toilet and biodigester and also by passengers' awareness.

A Memorandum of Understanding (MoU) was signed between the two government agencies on 9 March, 2010. A joint working group (JWG) comprising of scientists of Defence Research and Development Organization (DRDO), engineers of Rail Coach Factory (RCF), Integral Coach Factory (ICF), Research Designs and Standards Organization (RDSO) and administrative staff from Railway Board was constituted. The mandate of the JWG was to meet once in every three months for proper planning, execution, problem solving and rendering technical advice to the railway authorities on concerned matters. A mid-term review meeting was also planned at apex level involving senior officials from DRDO as well as Indian Railways to take stock of the implementation work.

During the course of implementation of Biodigesters in different trains, the major efforts have undergone for solving the problem of non-biodegradable material getting entry into the biodigester. Indian Railways provided the dustbins in the toilets of different trains and printed written instructions inside and outside of the toilets for creating awareness about the biodigester among passengers.

Indian Railways provided water seal through P-trap in all the toilets to stop foul smell emanating from biodigesters, if any. Further, different interfaces were designed to prevent the ingress of non-biodegradable material into the biodigester. It was decided to work on four options as given in the following table and to go for the trials in the passenger coaches.

Table 4.1: Comparative features of different types of biodigesters for resolving the issues related to ingress of non-biodegradable materials

Brief description	Features			
	Pneumatics	Electrics	PLC	Flush
System with flapper valve (Fig. 4.12)	Yes	Yes	Yes	Pressurized
System with manual slider valve (Fig. 4.13)	No	No	No	Gravity
System with reduced opening at inlet (Fig. 4.14)	No	No	No	Gravity
System with solid liquid separator (Fig. 4.15)	No	No	No	Gravity

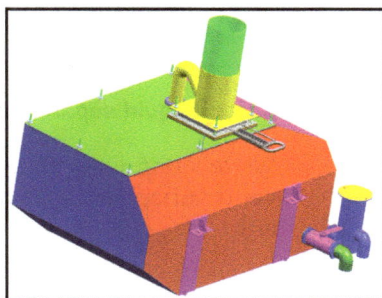

Fig. 4.12: PLC based flapper valve controlled biodigester

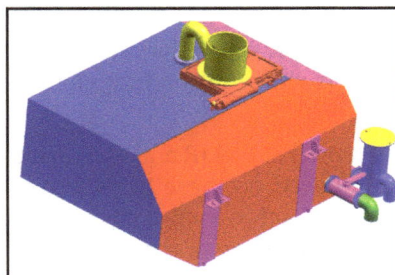

Fig. 4.13: Three dimensional diagram of Manual slider valve based biodigester

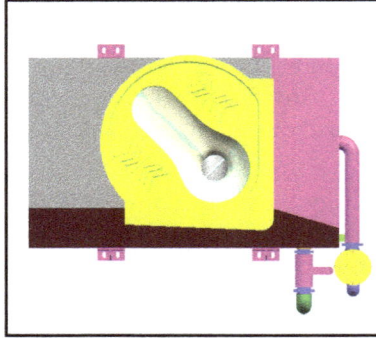

Fig. 4.14: Biodigester with reduced opening at inlet to restrict the entry of bottles, _etc._

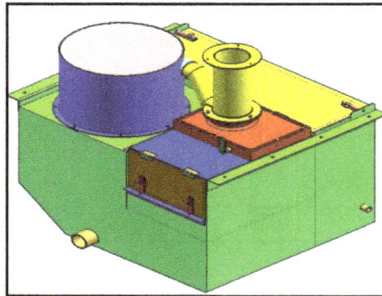

Fig. 4.15: Biodigester with solid liquid separation arrangement

The first variant had flapper valve along with pressurised flushing through pneumatic control whereas the second variant was provided with manual slider having normal gravity based flushing.

Another variant had narrow outlet from the toilet (50mm as against 150mm in conventional toilets) so that plastic bottles which were the main junk of the non-biodegradable material do not enter the main tank.

The fourth version had an additional cylindrical structure that separated all of the solids including human waste from the liquid. The liquid portion of the waste gets entry into the second chamber of the biodigester whereas solid are retained at the bottom of cylindrical structure in the first chamber itself for removal and subsequent treatment.

All these variants were fabricated by IR and were used in Bundelkhand Express for approximately one year. Based on these trials, options were narrowed down to manual slider valve system which was also subsequently changed to ball valve arrangement (FIg. 4.16) due to leakage and operational issues with slider valve system.

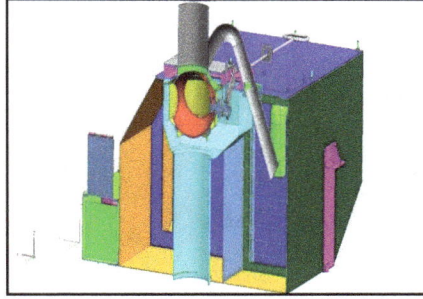

Fig. 4.16: Biodigester with ball valve arrangement for segregation of non-biodegradable waste

Satisfied with the performance of biodigester containing ball valve separator and P-trap, IR started trials of biodigesters at pilot scale in different trains in different zones. Joint Working Group (JWG) carried out the review of performance of final version of biodigester and recommended its implementation in all the passenger coaches of all the trains. Subsequently, at the instruction of Railway Board, Rail Coach Factory (RCF), Kapurthala, Integral Coach Factory (ICF), Chennai and various production units of railway undertook the responsibility of large scale implementation of biotoilets in passenger coaches.

5

Current Biodigester Technology for Indian Railways

Concerted efforts of scientists of DRDO and engineers of Indian Railways over a period of more than a decade led to the introduction of biodigester in passenger coaches. Rigorous field trials conducted by Indian Railway resulted in several modifications in the biodigester interface with toilet. Biodigester variants were installed in the passenger coaches and feedback obtained from passengers, cleaning staff and railway engineers were discussed and reviewed at different platforms both at working and decision making levels. Finally, decision was taken for installing the ball valve type of biodigester with P-trap arrangement in passenger coaches. Mounting of the biodigester beneath the coach has always been of great concern due to safety reasons in case of accidental fall of the equipment. Therefore, it was decided to install the biodigester on the under frame by positive mounting arrangement. Additionally, an iron rope was tied around the biodigester to hold it with coach in case of accidental fall. Suitable modifications were also made in the chlorination chamber as well as sampling port for increasing the retention time and getting the representative sample, respectively.

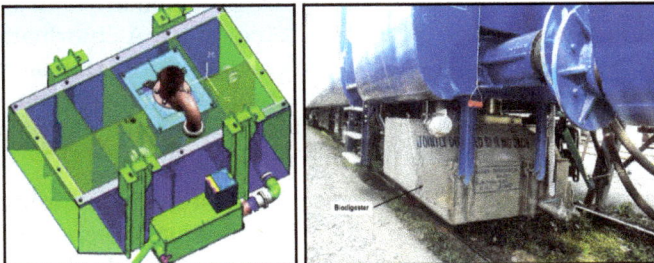

Fig. 5.1: Current biodigester operational in Indian Railways

The final variant opted for large scale implementation is shown in FIg. 5.1. The material selected for the final version of biodigester is stainless steel (SS316) with biodigester dimensions of 1150 mm (L) x 720 mm (W) x 540 mm (H) for treating the human excreta originating from single toilet. Biodigester weighs 110 kg, and has a total and working volumes of 400 L and 300 L, respectively. Various parts of biodigesters are shown in Fig. 5.2.

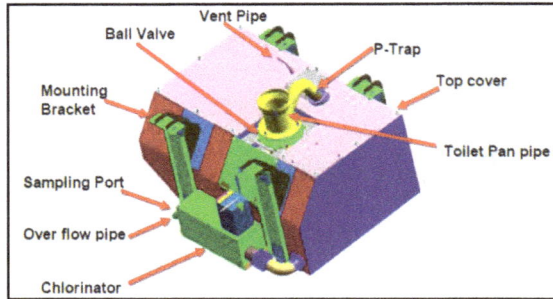

FIg. 5.2: Biodigester showing parts and their positions

The biodigester has six chambers separated by partition walls which have immobilization matrix pasted and fixed with iron mesh on both sides.

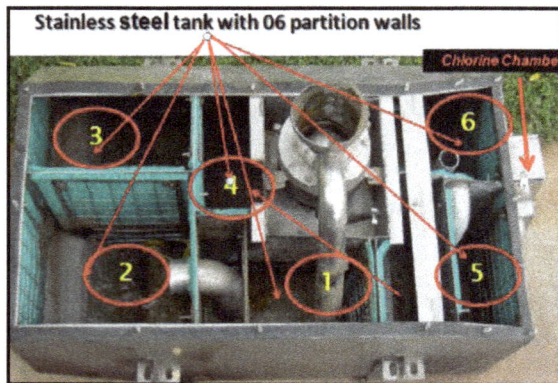

Fig. 5.3: Biodigester showing details of its partitions and other components

Openings in the partition walls (Fig. 5.3) are provided in a way to allow longer path for liquid flow in the tank till it reaches the outlet. A chlorinator having the provision to house chlorine tablets (75 mm diameter) is provided on exterior of the tank with the arrangement such that only one tablet at a time remains immersed in the effluent. A sample port is also provided in the chlorination chamber near the outlet for collecting samples for testing of effluent quality.

The human excreta from the toilet enters the first chamber through P-trap and travels through different chambers wherein fermentation of the waste is completed. Then the effluent enters to the chlorinator and gets disinfected before final discharge through outlet on the rail track. In case, any foreign material (non-biodegradable) is added to the toilet, it rests in the main pipe (before P-trap) and is thrown out when ball valve is operated by a lever arrangement. Pieces of pouches and sachets though may get entry into the first chamber through P-trap but their subsequent movement into the next chamber gets restricted by a wire mesh provided at the junction.

The start up of biodigester does not require the services of qualified personnel as the inoculum simply needs to be poured into the biodigester through the toilet pan and the biodigester is ready for use by passengers. The biodigester is charged with bacterial inoculum at a volume of 100 L per tank after its installation under the coach. It is desirable to use the toilet within few days (0-2 days) for bacteria to get activated, multiply and occupy full working volume of the tank for retaining desired anaerobic conditions. However, even if usage of toilets are delayed due to some reasons, the bacteria can still survive for weeks together inside the biodigesters.

The biodigester once inoculated and activated, continue working for the life-span subject to its continuous usage by passengers. The system is self sustainable as bacteria continue multiplying, forever, in presence of the food (human waste). The survival and the activity of bacteria do not necessitate any support like agitation and aeration as the system is anaerobic. Hence, there is no need of re-inoculating the tank with new bacteria anytime during its usage. There are no additional requirements for the optimum functioning of the biodigester except that toilet needs to be maintained in clean conditions as expected with other common toilets. However, the excess usage of washing water and cleaning agents should be avoided.

Disinfection of the effluent, discharged from the biodigester as a result of fermentation, is carried out before final discharge on the railway track for improving the quality in terms of pathogen inactivation besides odour and colour improvement. It is achieved by passing the effluent through a cylinder fitted outside the biodigester with a stack of chlorine tablets. Generally one chlorine tablet comes in contact with the effluent and dissolved chlorine being a strong oxidizing agent acts on any organic material present in the liquid including pathogens. It is the free chlorine, rather than bound form, that is responsible for pathogens inactivation. Hence, the quantity of chlorine required for disinfection depends on the organic matter present in the liquid. The availability of chlorine from the tablet has been considered based on these parameters. One tablet lasts for a period of 15-21 days depending on toilet usage and the stack of tablets need to be replenished on requirement basis.

The biodigester works on the principle of anaerobic bio-methanation of organic materials (human excreta). Human excreta is mainly composed of water, protein, starch, lipids and cellulose with protein as major organic component. The composition may vary with excreta of individual and cellulose is in higher quantity in case of vegetarians. The polymeric organic matter is hydrolysed by anaerobic bacteria producing extracellular hydrolytic enzymes and converted to monomers like glucose, amino acids, and fatty acids. Subsequently, another group of bacteria present in the inoculum convert them to low chain volatile fatty acids which further get converted sequentially to acetate, CO_2, hydrogen, etc. till methanogens convert them to methane (biogas).

Although, cold active bacteria work in wider temperature range and may have optimum temperature in mesophilic range but still are able to degrade the waste at temperature as low as 5 °C. This special feature of cold active bacteria, i.e., functioning in wide temperature range, enabled the biodigester to work in varied climatic conditions of the country wherein Indian Railways operates its trains round the clock, throughout the year.

The human excreta, while passing through different chambers of the biodigester, get hydrolysed and converted into biogas and water to an extent of more than 99%. Some of the leftover particulate material settles in the tank and gets degraded subsequently as it is exposed for longer treatment time. A little of organic material, that may get washed away with the effluent (contributing for BOD), hardly disturb the ecosystem as it is harmless and is composed of mainly ligno-cellulose.

Anaerobic conditions, reduced materials, fatty acids and other metabolites generated during fermentation inactivate the pathogens of human excreta to the extent of more than 99%. Moreover, most of the pathogens being aerobic or microaerophilic in nature, can not multiply and compete with anaerobic bacteria in strict anaerobic conditions even if they survive for short duration. The left over pathogens and off-odour (if any) are taken care by chlorine tablets when the effluent passes over it before getting discharged on the railway track. Arrangements have been made such that the effluent discharge from the biodigester does not fall on the rails but between the track to prevent corrosion. The corrosion of rail lines by human excreta has always been a big concern for the Railways authorities due to heavy maintenance and financial losses.

The biodigester jointly developed by DRDO and Indian Railways for on-board treatment of human waste in railway coaches has following inherent and practical advantages:

    i. The technology is scalable for implementation in the entire network of Indian Railways.

    ii. Being cost effective, it is implementable in the vast network of Indian Railways.

    iii. It does not involve high recurring cost for maintenance or replenishing the consumables.

    iv. The biodigester with defined volume of 300 Litres is appropriate for variable number of users ranging from 5-30 that may be the case as passenger number varies with the class of coach.

    v. The present arrangement of open discharge restricts the use of toilets by passengers at stations which otherwise is comfortable for the users. This technology does not impose any such restrictions.

    vi. It works in different geo-climatic conditions of the country wherever Indian Railways operates its passenger trains.

    vii. The biodigester works under fluctuating temperatures and is capable to withstand long duration (seasonal) as well as short duration (change in place) temperature variations.

viii. The technology works over a range of salt concentration in the water that is being available at stations across the country for filling the overhead tanks of the toilet and used for washing/ flushing.

ix. It permits the use of commercially available detergents and cleaning agents in justified quantity.

x. Biodigester Technology for human waste disposal is Green and Eco-friendly.

xi. The effluent is free from off odour, pathogens and perishable organic matter, and is, therefore, environment friendly.

xii. The bacterial inoculation is required only once during start up and thereafter bacterial population remains sustainable in the biodigester.

xiii. The technology is completely indigenous and none of its components as well as consumables need to be imported.

xiv. It is a scientific and practical solution for taking care of splashing of excreta on the walls of coaches and spreading of filth on railway tracks. It relieves the railway staff working in unhygienic and inhuman conditions.

xv. It is an innovative development and can be of great help in eliminating manual scavenging.

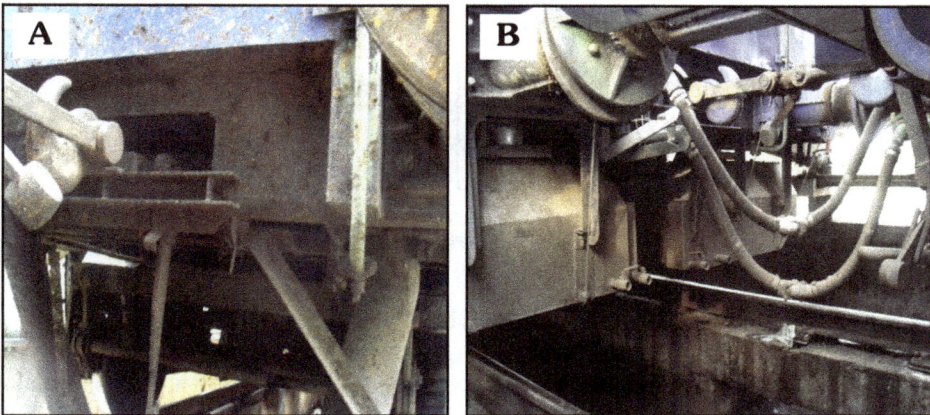

Fig. 5.4: Comparative view of filth on passenger coaches with biodigester (A) and without biodigester (B)

Implementation of Biodigester Technology in large number of trains, originating from different places and operating across the country, needs proper planning and systematic execution to help sustain the technology for the benefit of society.

Indian Railways, after struggling over a period of two decades to get rid of the stigma attached to open defecation in absence of a suitable technology, has now decided to install eco-friendly biodigester toilets in all its passenger coaches at the earliest possible. As per the terms and conditions of the Memorandum of Understanding signed between DRDO and Indian Railways in the year 2009 and renewed in 2014, Indian Railways is committed to implement the Biodigester Technology in all the passenger coaches. DRDO being the custodian of the Biodigester Technology and development partner of Indian Railways is continuously facilitating the implementation programme by addressing the troubleshoot from time to time and catering to the need of inoculum by itself or through its licensees. The quality control of inoculum as well as proper operation of inoculum generation plants at different places is also the responsibility of DRDO.

In order to maintain the regular availability of microbial inoculum, the mother bacterial culture (seed inoculum) is being propagated at DRDE Gwalior in 14 and 70 m³ inoculum generation facilities at defined temperature. DRDO has also supported Indian Railways in developing its own inoculum generation plant of 100 m³ capacity at Motibagh workshop in Nagpur. Besides, similar facilities have also come up in more than 18 industries (DRDO licensees) located in different regions of the country with a daily output of approximately 20000 L inoculum enough for 200 biodigesters per day. New industry partners and inoculum generation plants are being created to meet the growing demand of biodigesters in coming years.

Fig. 5.5: Underground anaerobic microbial inoculum (AMI) generation plant (70 m³ & 14 m³) at DRDE Gwalior

So far, the performance evaluation of biodigesters, operational in coaches, has been done by DRDE Gwalior and the samples are sent by different zonal railways to the laboratory at frequent intervals. Since, the scale of operation is increasing, day by day, across the country, it is neither recommended (due to distance) nor feasible for DRDE to analyse the required number of samples. Hence, Indian Railways has decided to establish quality control laboratories at their centres. Quality control laboratories have been established in the premises of Indian Railways at Nagpur, Gwalior, Delhi, Jaipur and other places.

A large industrial base has been created for fabrication of biodigester and production of bacterial inoculum by DRDO to help Indian Railways for implementing the technology. More than 55 industries have been provided with technical know-how by DRDO under a license agreement for the production of biodigester for mobile as well as stationary applications.

The bacterial inoculum generation facility is an important and critical infrastructure required for supply of seed material for biodigesters' start-up subsequent to their installation in coaches. Each biodigester requires 100 Litres of seed material for its proper functioning. Eighteen of the participating industries have functional plants in their premises with the total output capacity of approximately 7000000 Litres per annum. This quantity of inoculum is sufficient to inoculate 70000 biodigesters in a year that is more than the set target of biodigesters by the Indian Railways during the current financial year.

Indian Railways started installing biodigesters in its passenger coaches during 2012 and within one year took the decision to equip all its new coaches with the biodigester rolled out from Rail Coach Factory (RCF), Integral Coach Factory (ICF) and other production units with a commitment of fitting 16000 biodigesters under all the four toilet units of approximately 4000 passenger coaches manufactured every year. So far approximately 40000 biodigesters have been installed in different trains and a target of 30000 biodigesters has been fixed for the financial year 2016-17. Indian Railways has set a deadline of October 2019 to complete the installation of biotoilets in all the passenger coaches to commemorate the 150th birth anniversary of Mahatma Gandhi, the father of nation.

Approximately 40000 existing coaches, not fitted with biodigesters, are still operational in different trains of Indian Railways. Additionally, 4000 new coaches are inducted every year in the railway network. Thus, the requirement to retrofit these biodigesters on annual basis comes to approximately 72000 in 18000 passenger coaches per year to achieve the target date of October 2, 2019. Therefore, Indian Railways needs to revise its target substantially from the existing 30000 in the current as well as in coming years. Seeing the past track of induction of biodigesters since 2011, it is quite possible that Indian Railways will achieve the set target to provide healthy and hygienic environment to its passengers.

6

Imminent Social Impact of Biodigesters

Biodigesters fitted in the coaches of Indian Railways are poised to create environmental, social and economic impact within the country. Indian Railways traverse across the country and can be a role model for cleanliness for the masses. Direct impact of sanitation created by biodigester in short span of time will be visible on the Indian Railways with respect to its image, business and health of passengers as well as of people residing near railway tracks and railway stations. Being a competitive transport sector with roadways and airways, improved hygiene and sanitation is going to attract passengers for long and short distances not only of Indian origin but also foreigners who are conditioned for clean environment. It is a matter of dignity and pride for the railways as well as country while considering the foreign tourists as passengers. Clean environment will improve the railway's revenue in particular and country's economy in general.

Every day, 12617 passenger trains cover some part of 115000 km of track that includes 7112 stations. Between 2013 and 2014, Indian Railways transported over 8 billion passengers across the country, moving more people every day than residing in Taiwan. All passenger trains till recently had 'Open discharge' module toilets and about 3980 MT of human waste directly dumped onto the rail tracks polluting stations and the areas through which the trains pass including rivers and other water bodies. A chain of faecal pollution is formed and has been continued over almost a century. This chain of events is responsible for many health-related maledictions in our country as well as the poor health index of the country.

Open discharge of human excreta through 'chute' type of toilets of railways has a cascading effect of poor aesthetics, polluted water ways and disease spreading environment. Diseases associated with poor sanitation are well correlated with poverty and infantile deaths, and alone account for about 10% of the total disease burden. Exposure to human excreta is very dangerous to health as it is the store house of millions of pathogens both viral and bacterial that cause wide spread diseases. In many of the developing countries, at any point in time, around half of the hospital beds are occupied by people with diarrhoeal diseases. World Bank officials estimate that there are 4.5 lakh deaths out of 575 million cases of diarrhoea every year. It is reported that the prevention of sanitation- and water-related diseases could save Rupees 500 billion per year in health system alone. One can add another Rupees 400 billion per year towards averted man power losses. The World Bank report quantifies the economic losses to India, and shows that children and poor households bear the brunt of poor sanitation.

There cannot be any estimates in economic terms for the sufferings of individuals on health ground, but the losses with respect to man hours and treatment cost are assessable. Elimination of human waste, directly falling on railway track, employing innovative and maintenance-free technology, will have cascading effect on controlling diseases of faecal origin that affect passengers directly and the remaining society indirectly. Visible impact of biodigester technology on sanitation is bound to reach to whole society immediately and subsequently as passengers get educated and convinced to replicate it into their houses, colonies and public places.

It is known that sanitation has a direct bearing on health as well as to social and economic development of a country. Western world realised this fact quite early and toiled to make sanitation as one of the most important policy issues. They reaped the benefits from clean environment in terms of social and economic up-gradation of their countries. Somehow this has been overlooked for decades and lacked impetus from government and the masses alike in our country. Time is ripe enough for taking corrective actions to curtail the menace of poor sanitation. Effective sanitation provides barriers between excreta and humans in such a way as to break the disease transmission cycle.

Indian Railways initiated various options for improving sanitation status by making stations clean. One of the options was to clean the filth on Railway track by resorting to hose cleaning at stations. Dislodging the faecal matter from the track by washing only transfers waste from one place to another and finally making it to enter in to nearby water bodies. As water becoming scare, people are forced to consume this kind of polluted water and get sick.

It is estimated that Indian Railways spend on an average Rs 350 crore every year for the manual cleaning and waste collection from tracks. In cities, the Indian Railways contracts companies, many of which employ labours who are forced to do manual scavenging to clean stations, tracks and train toilets. But, there's no mechanism in place to clean the tracks as trains traverse vast expanses of uninhabited land. As per the white paper on Indian Railways published by the Ministry of Railways in 2015, one of the biggest challenges facing Indian Railways today is its inability to meet the demands of its passengers for cleanliness.

Although sanitation affects the human life and economy of the country in many different ways but the effect on individual's suffering specially the poor, children and women are of prime concern. Damages caused to the dignity of the country, environment and ecosystem are irreparable. These effects are bound to have short and long term impact on our image, economy and sustainability.

In spite of sanitation technology reaching such an advanced stage in modern times, lack of well formulated sanitation availability to masses is surprising. The first reason appears to be the stigma associated with it. The topic for discussion on toilets and treatment of human waste does not appear attractive and people simply feel uncomfortable. People also shy away to associate their names with sanitation activities while communicating with the society. Further, at first look, sanitation does not appear as urgent and essential as issue of access to drinking water or other matters, however, the sanitation and drinking water issues are closely related and equally demanding. Disease control is an impossible goal without proper sanitation adjustments. Besides the obvious health benefits, improved sanitation in developing countries would provide $ 9 economic benefit per $ 1 spent according to WHO.

There are a few key issues of sanitation: 1. Sanitation is vital for human health. 2. It affects poor the most. 3. Children and women are more susceptible. 4. It generates economic benefits. 5. Sanitation contributes to dignity and social development. 6. It helps maintain the environment. 7. It helps contributing for purer water sources. 8. Technologies and resources are available to tackle sanitation problems. 9. Awareness is the most important key for success. 10. It is achievable. In this era and on this date, no one should defecate publicly and no one should discharge untreated shit into the environment. No one has the right to risk and kill the others.

Tourism

Tourism in India is not only important with respect to economy of the country but it also gives the opportunity for showcasing our culture including life style, eating habits, history, art, architecture, etc to foreign tourists. Further, it has direct bearing on country's socio-economic structure in the form of economic gains and employment, besides exposing the people to the outside world to experience their culture and life style. It is important as most of the Indian people can not afford to go abroad to know rest of the world in terms of their eating habits, veshbhusha (dresses), languages, life style, culture, etc.

As per World Travel & Tourism Council's report, Indian tourism generated 8.31 lakh crore (US $ 120 billion) in 2015 which contributed for 6.3% of the nation's GDP. This sector is expected to grow at an average annual rate of 7.5% to 18.36 lakh crore (US $ 270 billion) by 2025 and will contribute to 7.2% of country's GDP. This sector has significant share in employment as well. The direct share of employment in tourism service industries is 4.4% and if indirect share is also included, this goes up to 10.2%. It means that almost every 4th person employed in non-agricultural activities is directly or indirectly engaged in tourism activities.

The number of foreign tourists travelling India in 1997 was 2.37 million that increased to 8.03 million by 2015. Although the number of travellers have increased over the years, the relative growth has dipped sharply and it is a matter of serious concern. Although some of this is due to an international slowdown in tourism in the aftermath of the economic crisis of 2008, there are plenty of home grown factors to blame for this plateau especially hygiene and sanitation.

Provision of good transport, affordable and suitable accommodation, safety of personnel and goods and maintenance of historical monuments are mandatory requirements for the tourists not only of foreign origin but domestic as well. Foreign tourists work as ambassadors of the country and whatever good or bad they experience, is communicated to rest of the world. Congenial environment will certainly enthuse others to visit the country and thus boost the tourism industry. The effect does not remain restricted to the specific industry alone, it boosts the collaboration and growth in diverse areas. Beside this, the important aspect that is mostly ignored and cannot be calculated in terms of monetary gains is the dignity and honour of the country. Indian culture believes on the concept of '*Atithi Devo Bhavai*' that means guests are equivalent to God. This principle makes them comfortable and happy that in turn will give big dividends automatically to the country without any desire or demand.

The most important aspect, linked to tourism, is maintaining cleanliness everywhere. Isn't it our duty to provide tourists with neat and clean environment, accommodation, transport and food? How it can be expected in the present sanitation scenario of the country when almost half of the country's population has no other choice rather than going for open defecation. The extent to which treatment of faecal waste in the existing toilets in India is being done is difficult to calculate but it is certainly very low. Open defecation in fields directly contaminates the fruits and vegetables as well as water sources and air. Rodents, house flies and other insects transmit the pathogens to other food sources. The difference in hygiene and sanitation status of the people from developed countries and developing countries does not need any documentary evidence and can be experienced in practical terms. If foreigners consume the food from common hotels/ restaurants or water from public supply, they are bound to get gastrointestinal infection. Tourists, visiting the common place including the market, cannot expect the availability of urinals or toilets and are forced to relieve themselves in open areas for the natural calls. This practice may be common for Indians but not for tourists from developed countries and it affects their psychology due to hardships faced by them. The country's negative image carried by them is unimaginable.

Biodigester Technology can play the key role in providing good hygiene and sanitation. This has been demonstrated and proved, over the years, in diverse geo-climatic conditions and situations in our country. Environment and user friendly biodigesters are unique solution to eradicate the problem of faecal discharge on railway tracks including railway stations, latter being more visible and significant as the place is used for halts and refreshments. These mobile toilets fitted in the coaches have shown visible impact all over the country and of late, few rail sectors have been declared as Green Corridors' where all of the passing trains have been equipped with biodigester based biotoilets. This helped in creating faecal- free routes and cleaner environment.

Indian Railways is committed to fit all its passenger coaches with biotoilets by October 2019. They are also expanding their wings to fit all the railway stations' public conveniences with biotoilets which are eco-friendly, hygienic, economic and self sustainable. Though, the railway is one among the several modes of transport to which people are exposed but it is certainly the most comfortable, economic and reliable mode of travel with wide connectivity. It is difficult to imagine the domestic or foreign tourism without railways as public transport system.

Biodigester technology has all desirable features and has proven potential to maintain hygiene and sanitation at stations and railway tracks spread throughout the country. Tourists can not go back to their respective countries without noticing the miraculous effects of this indigenous technology as most of the trains have been equipped with biodigesters. Days are not far off when this technology will find its application outside the country as well. Although, Indian Railways has taken initiative to export this technology to the developing world but meeting the domestic requirements over next three years gets the logical priority.

The technology has proven its versatility not only in Indian Railways operating its trains in different geo-climatic conditions across the country but also in rural and urban sectors in single houses to housing societies and public toilets. Government of India, state Governments and all stake holders of tourism industry are required to take initiatives for implementation of this fool proof innovative technology to each and every place where tourists are likely to visit, stay and shop.

Health

The poor access to water supply is a matter of concern worldwide with over 2.5 billion people having limited access to sanitation facilities. The global burden of disease and mortality rates could be reduced by about 9.1% and 6.3%, respectively, if success is attained on sanitation and hygiene facilities. A large proportion of these diseases are related to diarrhoea incidences which contribute to the mortality rate of about 1.9 million and new diarrhoea cases estimated at 4 billion annually especially among children under five years old. The World Health Statistics review done in 2009 showed that the highest case fatality rates with over 386000 deaths occurred in India due to diarrhoeal incidences.

Research suggests that lack of proper water, sanitation and waste management exposes many children to water-borne diseases. Besides their health effects, water-borne diseases also have adverse developmental consequences for children. In a study, a significant negative correlation was found between children's cognitive functioning and early childhood diarrhoea of 6.5 to 9 year-old children living in Brazil. The dumped garbage in the vicinity and access to toilet has been reported for differences in cognitive and psychomotor performance of low-income group 12-months old kids. Likewise, in an investigation of the environmental conditions including inadequate sewage drains impacted 7 to 8 years old children's cognitive development in war-torn Baghdad City. Higher rates of infection by helminths and protozoa were more prevalent in the younger age group consisting of children aged 7-8 years old compared to the older children. Studies have shown that children in grade 3 are at high risk of being infected with schistosome in communities with high prevalence of this disease.

Literature covers considerable studies regarding the effects of lack of appropriate water facilities, hand washing, and hygiene practices on child health outcomes. Impaired cognitive learning and learning performance are long-term outcomes of the negative effects of infections such as diarrhoea, worm infestations, and dehydrations which are largely attributed to poor water, sanitation, and hygiene conditions. Diarrheal incidences in children during their first few years of life have been shown to limit their growth by about 8 cm and cause an IQ point reduction when they progress to about 7 or 8 years of age.

In addition to direct impact on cognitive functioning, intestinal parasites and bacteria-contaminated water (often from sewage) contribute towards malnutrition and stunting, both of which impact children's IQ and school performance, and may also contribute towards behavioural problems. These effects may result as a consequence of early malnutrition and exposure to environmental toxins and stress. These factors can alter both brain structure and function, thus leading to long-term changes in cognitive and socio-emotional functioning. In addition, both illness and malnutrition may lead to increased absences from school and attention problems when in school.

Attendance is a strong predictive factor of academic success for elementary school pupil. Studies have shown that about 75% of all school absence are illness related. Information regarding absenteeism from middle and higher income countries has shown that poor academic and social development, high dropout rates, and reduced learning performance are attributed to school absence in children. Absenteeism due to illness has been shown to be reduced by implementation of mandatory hand hygiene and sanitary procedures based on the results of previous interventions. The availability and utilization of alcohol-based sanitizers in schools have also been shown to reduce absenteeism by about 20–50% .

There have been considerable studies that examined the effect of water treatment, hygiene, and sanitary practices on reducing absenteeism, diarrhoea prevalence, and acute respiratory infections in school-age children. However, limited research has been done to evaluate the effectiveness of water, sanitation, and hygiene practices through randomized controlled clinical trials to observe the long-term impact of these interventions on improving child health outcomes. All the studies done in developing countries reported a positive effect of sanitation practices on reducing diarrhoea prevalence.

Sanitation and hygiene are key to survival, development and growth of children. Improved sanitation is still a dream for people and has yet to reach to almost 980 million of children under 18 years in the developing world. The figures are not surprising and are shocking. Millions of children dying each year from preventable diseases are a black spot on modern world which boast of its scientific and technological achievements, industrial progress and luxurious life. Sanitation to all persons of the planet is feasible technically and viable economically but need prioritization, planning and determination. It is a basic need and cannot be ignored further for getting mentally and physically healthy future generations. The time for action is now.

Biodigester Technology has proven its worthiness in Indian Railways where sanitation had been a technologically difficult challenge due to mobile application, space constraint and environmental factors. On the contrary, application of biodigester in stationary mode at schools, houses and public places is much easier and economical. Expected impact of improved sanitation by such measures will have lower morbidity and mortality rates, better nutrition, better learning and retention among school children. It will provide dignity and privacy for everybody especially the girl students. Students will be the best messengers for spreading sanitation practices to their homes and societies on permanent basis. Therefore, sanitation practises in the schools and educating students in this area are of utmost importance.

Social Issues

When women do not have access to toilet, they are forced to go outside, which is hazardous in broad perspective. This also means travelling long distances, often at night, in order to retain privacy and dignity under the cover of darkness. This increases the risk of harassment, sexual violence and even rape which are very serious issues that must be addressed.

Keeping women in the work place with poor sanitation has a domino effect. Considering the impact of unhygienic sanitation facilities on health, lack of facilities in the workplace can have an impact on absenteeism, affecting livelihoods, productivity levels and ultimately the economy. And as women go through different stages of life including pregnancy and old age besides temporary or permanent disability, the proximity of the toilet becomes all the more important.

Lack of adequate sanitation affects everyone but women and children often bear the brunt of the lack of toilets and other sanitation facilities. Not only do women and girls have different physical needs but they also have greater need for privacy while using toilets and bathing. Inaccessible toilets and bathrooms make them more vulnerable to physical violence and other forms of gender-based violence. Women and girls, more than men, suffer the indignity of being forced to defecate and urinate in the open. They disproportionately face risks of sexual violence even when they walk, on average, more than 300 metres from their homes to use available latrines or open fields.

In the absence of sanitary facilities, women often have to wait until dark to go for toilet located distantly or otherwise in open field. To delay urination, women often drink less water leading to a risk of health problems. Attempting to hold bladder until evening may result in physical harm beside urinary tract infections. People may also attempt to eat less or modify their diets, by not eating certain fibrous foods such as pulses or leafy vegetables to delay the emptying of stomach. An unbalanced diet may result in negative effect on health especially with long-term consequences.

In addition to the risk of physical and sexual violence, women and girls who defecate in the open, especially in the bush, face the risk of animal attacks. Women and girls are more susceptible to snake bite because they tend to move quietly in the bush in order to remain hidden. Snakes and other animals are then not scared away and are more likely to be surprised by the women's presence and bite them. Men, on the other hand, are more likely to walk loudly into the bush scaring snakes and other animals away.

Women and girls don't need toilets and bathrooms just for defecation. They also have a much greater need for privacy and dignity when menstruating. Women and girls have particular sanitation needs when they are menstruating which are rarely discussed and considered. Menstruation remains a taboo in most of the cultures. It impacts their day-to-day working including employment, education in school or college, movement and other activities especially when they lack access to an appropriate sanitation facility. Separate toilets in school, for example, mean more girls are likely to join the school in the first standard, and more girls are likely to stay after puberty to complete their education.

Although women and girls place higher value on the need for a toilet than men but they are rarely in control of the household budget, and therefore access to sanitation remains a low priority in many parts of the world. Lack of space for toilets and waste disposal in low income group houses in congested colonies, high cost, cleaning needs, maintenance and off odour of waste/ effluent of treatment system, if available, also dissuade the family members to construct the toilets in close vicinity of the houses. DRDO biodigester based toilets address all these issues and people can be easily persuaded for access of toilets in the houses.

In most societies, women have the primary responsibility for the management of household water supply, sanitation and health. Water is necessary not only for drinking, but also for food preparation, care of domestic animals, personal hygiene, cleaning, washing and waste disposal. In rural areas, water is not always available at each door step. It is the responsibility of women to collect the water from out door, sometimes available at a distance of many kilometres. Requirement of less water for biodigester based toilets will reduce their hardships and save their energy, efforts and time to a significant level to be utilised in other essential activities.

While construction and maintenance of pit latrines (digging, repairing and exhausting) and other versions of toilets are primarily carried out by men, cleaning of *toilets* is primarily the responsibility of women. However, in some regions, the task of emptying the latrines falls exclusively on the shoulders of poor women, and the labour-conditions under which they do this work are inhumane. Biodigester is a boon to ladies as it does not require any cleaning and maintenance throughout its life of operation.

Environmental Issues

There are several countries in the world whose economy is heavily dependent on aquatic life mainly fishes, prawns, crabs, etc. In India itself, people of some states like West Bengal, Odisha, Kerala, Assam and other north-east states can not think their living in the absence of fishes. The middle and higher income group people can afford buying the fishes just like vegetables and their origin may be from sea or pisciculture farms but poor people are totally dependent on local water bodies like ponds, rivers and back water for harvesting these aquatic fauna for daily needs. Contamination of these water bodies with untreated human waste/ sewage is not only detrimental for these creatures by depleting dissolved oxygen in the water but also affect the human health by transmitting pathogens and other toxins through them.

Once the human waste gets entry into water bodies, it increases organic load of the water in terms of biological oxygen demand (BOD) and chemical oxygen demand (COD). High organic load causes the depletion of dissolved oxygen in water. A particular level of oxygen is essential for aquatic animals like fishes that are solely dependent for their survival on dissolved oxygen of the water which they consume through gills. Depletion of oxygen below five ppm leads to the death of such aquatic animals. Further, dissolved oxygen is also required for the survival of aquatic flora. Hence contamination of organic matter in the water bodies leads to disturbance of aquatic ecosystem.

High organic content in the water bodies leads to partial anaerobic conditions and consequent biodegradation generates lot of sulphides, H_2S and other toxic materials. It is a general practice to use such water for irrigation purposes for staple crops and mostly for cultivation of vegetables. Some farmers prefer such waste water for irrigation as it meets the additional requirement of nutrients for the crop. This can be sighted on the banks of Yamuna in Delhi and other cities where ever the river flows. The toxic materials of polluted water may adversely affect the vegetation and human/ animal health following consumption of the contaminated eatables.

Psychological Health

Biodigester (biotoilet) being an on-site sanitation unit eliminates the need for the disposal of human waste. The system is sustainable, totally eco-friendly, conserves water, produces fuel gas and never to be emptied. It is known that serene environment stimulates positive feelings and elevates mood. There are a number of studies which correlate the cleanliness and psychological health. Dr. Simone Schnall of the University of Plymouth and her colleagues linked physical cleanliness with a sense of moral duty. She found "that when feelings of disgust against filth are instilled in them beforehand, people make decisions which are more ethical than would otherwise be expected." This means that cleaning generates a feeling of moral worth in people, in the same vein as helping an elderly person across a busy street or giving money to a homeless person. The experimentalists concluded that elevation of prosocial behaviour motivates altruism, thus potentially providing an avenue for increasing the general level of prosociality in society. This elevation may lead to positive behavioural change. Therefore, cleanliness inducted through absence of open discharge of faecal matter will give rise to enormous social benefits. Travel is a kind of education and if this education has sanitation angle by witnessing pleasant surroundings, one can get motivated to contribute for cleaner surroundings.

Manual Scavenging

India is proud of its rich culture, great religion, inclusive language, diversity of traditions and kind hearted people but cursed due to its caste system. Division of labour led to four castes: Brahmana, Kshatriya, Vaishya and Shudra. Majority of people belonging to Shudras were assigned the task of manual scavenging. The nature of work, slowly, marginalized the group and they suffered on both fronts, i.e., socially and economically. In the modern era, lots of initiatives have been taken by Govt. as well as other non-governmental organizations to bridge this gap and success has been achieved to a large extent. Eradication of manual scavenging was taken seriously by enacting "The Employment of Manual Scavengers and Construction of Dry Latrines (Prohibition) Act, 1993" and " The Prohibition of Employment as Manual Scavengers and their Rehabilitation Act 2013".

Inspite of all initiatives and seriousness, manual scavenging is still practised, may be to a lesser extent, by only those groups of people who were doing this task since ages. The main reason for this is the absence of suitable sanitation technologies where scientists and technologists intervention is mandatory on account of moral and social responsibility. The central sewage treatment systems which are supposed to be automated and free from major manual intervention need the cleaning of sewage pipelines by the people of the suffering community. There have been frequent incidences when these people lose their lives while cleaning sewerage pipes because of the toxic gases like carbon monoxide and hydrogen sulphide.

Decentralized systems were practiced since ages and required manual scavenging periodically. Lot of innovations have undertaken on improvement of these systems and now better toilets comprising of septic tanks, twin pits, dry pits, etc. are available. However, these all suffer with the drawback of cleaning the residues at certain intervals and can not free the specific group of people from the task of scavenging. Biodigester Technology has nicely addressed this problem as the tank does not require any cleaning throughout its life.

After acquiring a technology that takes care of the menace of direct discharge of human faeces on to the railway tract, Indian Railways now can play a leading role in inculcating cleanliness culture in masses. It can show case the effects of clean ambience and in turn can be an important change agent for sensitizing people towards better sanitation. Trains of Indian Railways are not just the means of transport in our country but can be considered as the window of education through which one is exposed to vast cultural diversity, geographically varied terrains, imposing forts, majestic temples, etc. It is the medium through which one interacts with people of different languages and dialects. It enlightens one to the cultural values and cuisines across the country. Therefore if one experiences cleanliness at group of stations, in one part of the country while travelling, he will expect the whole country to be clean. This expectation regarding cleanliness will goad railways and passengers alike to maintain surrounding clean.

Indian Railways can take pride in claiming to have the most effective, economical and State-of-the-Art technology for the on-board degradation of human waste by adopting Biodigester Technology developed by DRDO. As per available literature, currently this kind of technology is not being utilised elsewhere in the world. Since the Biodigester is clean as well as green technology it will go a long way in improving the cleanliness status of stations and the tracks throughout the length and width of the country.

Considering the vastness and the number of people utilising the railways, the sanitation message one will draw from the cleaner rail tracks, will go a long way in inculcating a range of implicit habits beneficial to the society and the nation alike. Implicit financial gain from the technology of biodigester by Railways can be considered by knowing the axiomatic evidences across the world from clean environments like elimination of water borne diseases, healthy growth of children, absence of manpower losses and higher productivity.

The new scenario (clean stations and tracks) will have a cascading effect on social factors as well, like sense of pride, sense of well being and clean surroundings which in turn will improve tourism. Cleaner environment will have explicit effects on tourism and improved health index of the country and implicit implications like elevated sense of well being and resultant motivational urge to contribute for the cleanliness.

Sooner the diffusion of this innovation takes place in our country, brighter are the chances for the clean and healthy environment to live with. In order to achieve sanitation goals set by the Govt. of India, cleanliness affected by the biodigester technology alone may not be sufficient, and to ensure sanitation adoption becomes a way of life a "carrot and stick" approach may be needed. Also sanitation coverage can be increased through a combination of community-based promotion and enforcement of national or local legislation.

7

Future Endeavours

Every accepted technology undergoes a Technology Development Cycle consisting of five stages namely Research and development, Scientific demonstration, System deployment, Diffusion and Commercial maturity. The life span of a technology depends on these factors also. The present DRDO technology has travelled through the first four stages successfully and is in the fifth stage now. Usually some technologies like steel or cement manufacturing have a long life span because they are developed after evaluating detailed chemical engineering principles and processes. DRDO developed biodigester technology also falls under this category where it will continue to remain viable for years to come notwithstanding newer innovations. This optimism is based on rigorous and protracted trials that have undergone in different terrains and climatic conditions. In addition, the technology has been built on sound biotechnological principles.

The technology-life-cycle (TLC) for electronic and consumer-goods-technology is normally short due to heavy competition as the business is driven mainly through financial gains. Since the present technology has been implanted for a social cause, there is little room for the 'declining phase' as it does not entail explicit financial gains. More importantly the biodigester technology falls under 'disruptive-technology' as it has 'outsmart' costly foreign technologies. More recent sources also include "significant societal impact" as an aspect of disruptive innovation.

In spite of the success of this technology, development of a competitive alternate (improved) product or process has to be visualized to improve the lifespan of the present technology. As a result, continual efforts are on at DRDE to enhance the 'technology-life-cycle' (TLC) of Biodigester Technology by introducing innovative steps.

Present Scenario

A sustainable Biodigester Technology developed by Defence R&D Organisation (DRDO) for the onsite treatment of human waste has been seamlessly coupled in the coaches of Indian Railways (IR). These two Govt organisations of India, have provided an economical and immediate solution to the problem of open discharge of faecal matter on the railway tracks for which a workable solution had been sought over decades. Sustainable sanitation implies that the approach to achieve sanitation must be socially acceptable and economically attractive. The technology that has gone into the service of railway passengers fulfils these criteria very easily as this has been tested on-board over about six years.

With easy fitment, low operational cost and fool-proof functional features, there appears to be no reason to doubt the claims of Indian Railways that in the next 3-4 years, all of its coaches will be equipped with DRDO developed biodigesters. The reasons for such an optimistic target is that the decision has been taken after understanding the full potential of the technology, thorough testing and evaluation in laboratory, field as well as on-board-trials. It also involved several deliberations and threadbare analysis besides engineering modifications in a continuous manner over a period of one decade.

The Railways have tried and tested different technologies from overseas for the on-board-disposal/ treatment of human waste for the past many years. Though, the earlier adopted technologies proved futile under operating conditions in India, the failure of these foreign technologies came as a blessing in disguise for the country in general and Indian Railways in particular to test an indigenous technology which had matured within the country at the appropriate time for similar purpose.

The biodigester technology has been adopted in various short (less than one day) and long distance trains (3-4 days) to evaluate the faecal load vis-a-vis the capacity of the biodigester tank. They were also fitted in different type of coaches i.e. General, Sleeper, AC and chair cars to cover the range of utility requirement and attitudinal aspect of passengers. Moreover, these trains operate through different climatic zones - some in low temperature (during winters), others at high temperature of 40-48 °C (during summer). There are trains that operate through high and low temperature climate in the same journey while crossing southern and northern regions during winters. During the last 6-7 years of experience, authorities and staff of IR have gained confidence that biodigester performs well in all long and short variants of train journey and in different temperature zones.

All new coaches are being fitted with biodigester not only in ICF and RCF but in other Production Units across the country and the work is on at war footing for the retro fitment of old coaches.

In order to address the issues of dumping non biodegradable unwanted material (bottles, napkins etc.) into the toilet pan by the 'passengers' suitable provisions in the design of biodigester have been made by the Indian Railways. Added to this, the increased awareness and response of using bio- toilet by passenger has improved the situation significantly.

All technologies have to be upgraded continuously to prolong the "vital life" of the product cycle. This is best achieved by improving user 'satisfaction' through innovative thinking to enhance the capability of the technology.

A successful endeavour like biodigester, gives satisfaction to scientists who have toiled to make a process into a usable product for its mass usage. Implementers also get the pleasure of providing the solution to the problems, especially if they had been struggling for a long time. It is said that no one should rejoice the laurels for long. The success should give partial and temporary respite and one should not yield into comfort zone which mostly results into end of the efforts. There remains always a lot of scope to improve the technology for making it more environment and user friendly. Economics and various practical aspects are to be continually improved besides making the technology to face any competitions in the future which is more likely in the product of mass usage.

Future Trends

Since DRDO has waded all through on this technology from project conceptualization, stages of development of prototypes to a usable technology, futuristic road map is being contemplated continuously. The developing team has projected the following domains for further improvement of the technology in coming years:

- Material of fabrication
- Weight reduction
- Better efficiency
- Water conservation
- Harvesting biogas
- Better effluent
- Inoculum quantity
- Education and awareness

Material of Fabrication

Presently, biodigesters for railways are being made of stainless steel. It has good strength, rust resistance, malleability and durability. However, there is scope of alternative fabrication material which is lighter, have higher strength and longer endurance besides being cost effective. The use of composite materials has become an increasingly important factor in engineering design. Engineered plastic composites are gradually finding scope for wider applications in industries. Aerospace industry has been using Fibre reinforced plastics (FRP), because of their excellent mechanical and load bearing properties. Their intrinsic properties lend them to easy adaptation and effective customization.

A variety of plastics and their composite materials are available in the market and needs to be explored for the construction of biodigester tanks. Glass fibre reinforced plastic material has been used previously in fabrication of biodigesters in one of the earlier trials of IR by one of the industry (DRDO industry partner). However, a comparative evaluation has to be made between different materials in terms of weight reduction, mechanical properties including compressive strength and flexural endurance *vis-a-vis* economics of fabrication. Reinforcement is possible with many other synthetic and natural materials to get the desired properties in the fabrication material. Alternative metals and alloys also need to be thought of and worked out besides exploiting the nano-based composites as structural materials.

Existing biodigester weighs approximately 410 kg including its fixtures and liquid content. It is fixed below the coach like other fitments. Stainless Steel provides stable fitment, sufficient mechanical strength and impact resistance so as to tolerate the ballast hits by gravels stones (laid on rail tracks).

These stainless steel biodigesters, though, have required strength for the purpose of its objective but may turn to be a risk for speedy train in case of faulty fitment or during unfortunate train accident. Though it applies equally to other fixtures/ components below the coach as well, but selecting a material for biodigester with workable strength on one hand and good fragility on the other can certainly be thought of, explored and debated in future.

Weight Reduction

Rising energy prices and environmental issues are the global concern that demand the better performance from materials being used in railway carriages. 'Lighter the better' is the 'mantra' for reducing capital and operational cost and improving the performance of Railways.

Each bogie with a fitment of four biodigesters adds approximately 2 tonnes of weight leading to overall load in a train to the tune of about 48 tonnes. Thus, the significant increase in the load of a train is bound to increase the consumption of fuel as well as wear and tear of wheels and rail track. Even a small quantum of weight reduction will have significant impact on operational cost of railways.

Use of light weight fabrication material may reduce the weight of a biodigester by 50 kg and of 4 numbers by 200 kg. Light weight material having sufficient strength and other desired features may be explored by scientists and engineers and trials conducted considering all safety features.

The other possibility of reducing the weight is by decreasing the size/ volume of biodigester. There is a scope for reduction of Head Space which is provided for gases, froths or splashes. It is expected to reduce the volume of tank and weight by 10% comfortably.

Another alternative is to increase the fermentation efficiency and thus lowering the hydraulic retention time (HRT). Lower HRT means that same quantity of liquid waste is fermented in lower capacity vessel.

Yet other attempts can be diverted towards combining the waste of two adjoining toilets or all four toilets in a coach to carry out the treatment. It is also possible to collect the waste of all toilets of a train at one common place (preferably at the end of the train). The required volume of biodigesters for such options will not be simple arithmetic sum but will be much lower and volume saved is expected to be in proportion to number of toilets whose waste have been combined. Any of the combinations may reduce the volume requirement by 10-20%.

There are other assemblies fitted with the biodigester like chlorination and ball valve. Chlorination assembly can be made of light weight plastic and of smaller size and need not have excess water holding capacity. Suitable improvement in bacterial inoculum, e.g., by incorporation of specific phages for bacterial pathogens may eliminate the need of chlorination however, viability of constituent bacteria in the inoculum as well as its performance (especially hydrolysis) should not be compromised.

P-trap, a water seal, is also provided to stop any bad smell emanating from biodigester to enter into the toilet. Inflammable biogas produced in the biodigester also gets stopped from entering into the toilet by the same device. Submerged inlet, as has been provided in biodigesters designed for stationary applications to prevent the entry of foul smell from head space of biodigester into the toilet, may also be adapted to biodigesters of mobile applications, thus, completely eliminating the need of P-trap. Moreover, with P-trap larger quantity of water is required for flushing the toilet. Indian Railways has experience of pressurized flushing/ vacuum toilets and such systems are in operation in some of the toilets. Thus pressurized flushing/ vacuum toilets if attached to biodigesters, water requirement will be reduced to at least one half leading to need of lowered volume biodigesters.

The ball valve has been provided to stop the non-biodegradable ingress into the biodigester. By way of educating the commuters and by providing the other arrangements, it is possible to eliminate these extra fixtures of the biodigester. Ball valve is an external assembly and fitted before 'P' trap to stop the entry of non-biodegradable materials like plastic bottles and bags, sanitary pads, etc. into the biodigester and removing them as per the necessity. Presently such materials are being discarded by passengers into the toilet pan as the toilet outlet is open and the material directly falls on the rail track. Thus the mindset of discarding materials has to be changed with respect to biotoilets. Provision of dust bin in the toilet and awareness among passengers about these intricacies will certainly help to resolve such minor issues and IR has already taken such initiatives. Simple modifications in toilet pan (smaller size) and outgoing pipe (hooks, smaller diameter) along with other measures will certainly eliminate the requirement of such devices thus saving the weight requirement and the total cost. Further, the ball valve arrangement also necessitates the repeated maintenance of device thus further burdening IR towards finance although to a little extent but in efforts to a significant way.

Better Efficiency

Efficiency of biodigester is the most important aspect of the technology and can be improved in different ways. There can be no end to improve the inoculum in terms of biodegradation efficacy. Biodegradability efficiency of the inoculum can be improved further by different scientific measures. Bio-augmentation of bacterial culture can be a continuous effort by scientists which is possible by isolating higher efficiency anaerobic bacteria from the environment or other anaerobic human/

sewage treatment plant operating in different regions. Supplementing the bacterial culture with efficient methanogens will be of great help as they ultimately control the total biodegradation process of organic waste. Human waste, especially of person having vegetarian diet (common in India) has significant proportion of cellulosic fibres. High performing cellulolytic bacteria also need to be present in good numbers in biodigester which otherwise get compromised in protein rich environment especially at lower HRT. These bacteria can also be selected from cow dung biogas plants.

Genetically modified bacteria may find application in biodigester inoculum in 'future endeavours' as the mechanism of genetic manipulation is well known to the scientists and many such microbes are being used in the biotechnology-based industries. It is possible to introduce the genes of different hydrolytic bacteria into single bacterium. Besides, efforts can be made to increase the efficiency and workability in wider temperature range and odd situations. However, use of such microbes need regulatory clearances from concerned authorities since they will be released into the environment along with the effluent generated on-board in passenger coaches.

Efficiency of biodegradation can also be further improved by providing longer path for waste in the tank, by retaining higher bacterial mass and settling of suspended waste for attaining longer retention time for the undigested organic matter. Suitable modifications in design of the main tank to cater these requirements and subsequent field trials may be worth exerting. Further, efforts can be made to immobilize higher numbers of useful bacteria on alternative matrices having better affinity for bacteria, larger surface area and durability. Although, the presently used PVC matrix has been chosen after screening large number of matrices but probability of getting better alternative having low cost or otherwise can not be ignored.

Any biochemical reaction is said to get doubled with every 10 °C rise in the temperature and same is true for the bacteria within their working temperature range. Trains which are operated in low temperature regions can be provided with heating-element cum-sensor devices that maintains a required temperature of biodigester for efficient functioning. This heating system along with thermostat may also be provided in trains which operate partly in low temperature areas.

Water Conservation

Water is very precious commodity in today's world and needs to be judiciously used and conserved. The need of water for microbial activity in biodigester per person is not more than 250 ml but water is being used in quantities of more than 2 L per person to clean the toilet. As discussed earlier, the requirement of water in the toilet can easily be reduced by using pressurized water for cleaning the toilets and elimination of P-trap. Additionally or alternatively, effluent generated by the biodigester can be used for flushing the toilet. Effluent of biodigester is tolerable in terms of odour, colour, pathogens and organic content for flushing the toilet pan. However, further finishing of effluent water, if desired, is possible by sedimentation of suspended solids in the last compartment of fermentation tank and/ or in the storage tank at the roof of the coach. In addition, further augmentation of the colour and odour of effluent is possible by permitted synthetic/ natural additives. The recirculation of effluent will increase the number of useful bacteria in the digestion tank which is bound to increase the biodegradation efficiency of the human waste.

Any measure taken to reduce the quantity of water used in the toilet will have added advantage of higher biodegradation of waste as organic matter will be held for longer duration in the biodigester (higher HRT) or alternatively the size of biodigester will be significantly reduced in case the same HRT is maintained.

Harvesting Biogas

Anaerobic fermentation of human waste in the biodigester generates biogas and the quantity and the methane content varies from 5-10L per person per day and 55-65%, respectively depending on the temperature, water usage, cleaning agents used and journey duration. Methane is a well known green house gas and its release into the environment needs to be relooked. The matter is not very serious as the quantity of methane produced is negligible if compared with cattles (few cattles in a day produce methane equivalent to a full train in a year). Further, methane is also produced in enormous quantities naturally from paddy fields, organic waste dump, decaying agriculture and forest residues and even by human beings and other animals.

In spite of it, the biogas generated in the coaches can be used as energy source as a spin off benefit during the disposal of human waste. Biogas through pipes can be collected at one place and used for cooking purpose in pantry car provided in most of the trains. Otherwise, it can also be collected in balloons or rubber bags and used to generate the power in same train or send to other place of use.

Better Effluent

Central Pollution Control Board (CPCB) has made effluent discharge standards but they are primarily meant for centralized waste treatment systems like STPs for municipalities or industrial wastes. The standards for decentralized systems as for septic tanks and individual toilets have not been created at national and international level. One possible reason may be that present single step biodegradation process practiced in individual houses, community biotoilets or septic tanks can not generate the effluent of the desired quality as expected by regulatory agencies. Moreover, its monitoring at individual level by authorities is also not an easy task. The developing countries like India where open defecation, in absence of toilets, accounts to almost 50% of the population, the treatment and effluent standards get back seat.

In spite of limitations of decentralized waste treatment systems and absence of effluent standards, there is a dire need to develop the standards for monitoring the human waste treatment technologies so that our pristine environment is not spoiled. These standards need to be evolved with the development of technologies and time. Initially, the standards may be relaxed but to be stringent in future as technology gets upgraded and confidence is built-up.

The existing biodigesters mostly meet the requirements of effluent quality standards jointly set by IR and DRDO. However, further improvements can be made by gradually increasing the fermentation efficiency. More efficient technology will allow more stringent effluent standards to be practiced in future. The visualized efforts, as described above are enumerated as improved anaerobic bacteria inoculum, retention of higher bacterial mass in the tank, alternative immobilization matrix, improved tank design, reduced water quantity, etc. Most of the effluent BOD in present system accounts to settleable rather than soluble organic matter which can be easily taken care by providing sedimentation arrangements in last chamber of biodigester. Beside these, alternatives to chlorine tablets for disinfection as well as colour and odour improvement can be worked upon and some work in this direction by making use of potassium permanganate tablets is underway at DRDE, Gwalior.

Coliphages and other pathogenic bacteriophages in the inoculum may substantially inactivate/ kill the residual pathogens in the biodigester which has been well proven in our laboratory experiments. However, their presence may influence the fermentation of organic waste, differently at different temperatures which is yet to be studied. This aspect is very critical before taking any final decision, otherwise, basic purpose of technology gets defeated. The multiplying aspect of phages, whether with bacterial inoculum or independently, can be considered subsequently along with their survivability duration.

Inoculum Quantity

Biodigesters are initially charged with bacterium inoculum at the rate of 30% (v/v) and is needed only once throughout the life cycle of the biodigester. Looking the number of trains in India and biodigester requirement, large quantity of seeding material is needed. This necessitates large infrastructure for its production and initial as well recurring cost at the end of industry. Further, it has burden for transportation from place of production to place of use. Scientists at DRDE are exploring various options to reduce the requirement of inoculum, in first phase to 20% and finally to 10%.

Experiments are targeted to increase the bacterial density in inoculum by fermentation/ multiplication with reduced water quantity, by separating water/ liquid and by growing bacteria on solid high surface matrix which can be easily separated for subsequent use. Alternatively, minimum quantity of seed can be further multiplied in the biodigester itself in approximate period of 2 days before use of the toilets with the help of reducing chemicals and nutrients.

Beside this, efforts are worth-exerting to develop suitable one time usable/ long duration flexible/ rigid containers for transportation of inoculum for long and short distances. They should be developed considering the gas generation during transit and economics of production. Work is in progress to find out the chemicals that interfere in the metabolism of methanogenic bacteria so that gas production stops during transit without compromising the viability of useful bacteria.

Education and Awareness

Any excellent technology may have all advantages but if not practiced and appreciated by the users for which it is made is not of any use. It takes quite a long time for consumers to accept and make use of any new product. It is because of mindset, awareness and confidence generated with existing product. So far, toilets in the coaches had open chute wherein the waste was discharged directly on to the track and these chutes were also being used for throwing the other bio-degradable and non-biodegradable waste. Many of the passengers are still not aware that open chute has been replaced with closed systems like biodigester and they continue old practice of throwing the garbage. Others, in spite of knowing about new arrangement, carelessly continue the old practice. Awareness among passengers will not only educate them about the new technology but also sensitize about the importance of the new system for application in other places as well.

Fig. 7.1: Educative slogan about biodigester based toilet

Already, railways have made efforts in this direction and slogans are written on the door and inside the toilet for not throwing the waste materials in the biotoilet, besides, displaying the awareness programmes about biotoilet on the railway platforms through audio-visual systems. But the efforts are still not enough. More efforts are required through posters, audio-visual aids, exhibits and magazines in addition to personal communications. Besides the soft measures, punitive measures are also required to be in place to deter the offenders. Such measures will be binding to passengers both morally and legally to make proper use of the biotoilets. Seeing the inherent advantages of the Biodigester Technology coupled with passenger acceptance, days are not far off when railway stations of India will become the model stations with reference to sanitation even for the international community.

Fig. 1.1: Garbage around the railway track (p. 3)

Fig. 1.2: Cleaning of garbage by water from the railway track at the platform (p. 4)

Fig. 1.3: Removal of human excreta by manual scavenging (p. 5)

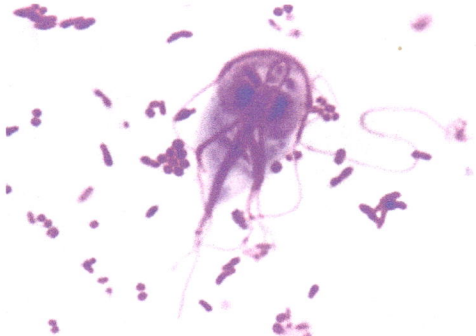

Fig. 2.2: Giardia (p. 15)

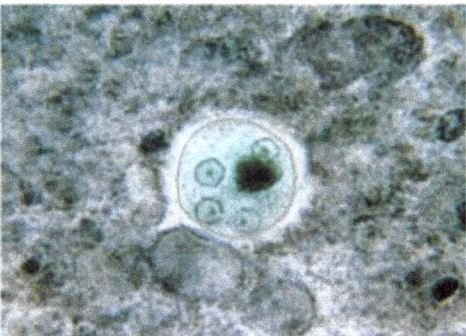

Fig. 2.3: Entamoeba histolytica (p. 16)

Fig. 3.1: LHB coach fitted with controlled discharge toilet unit (p. 22)

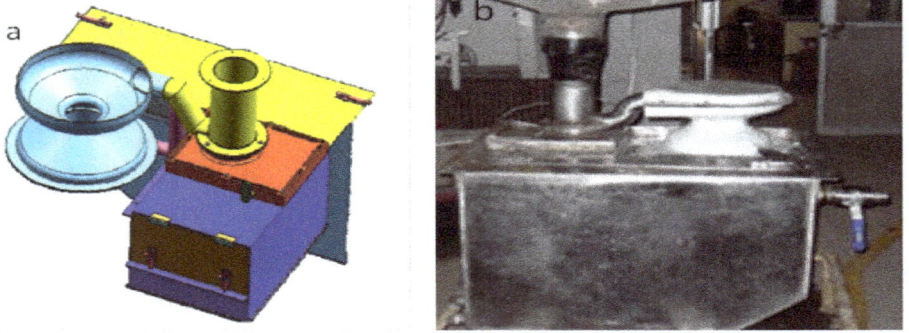

Fig. 3.2: Zero discharge toilet system model (a) and prototype (b) (p. 25)

Fig. 4.1: Temperature controlled biodigester in snow bound areas (p. 32)

Fig. 4.3: PVC matrix for immobilization of bacteria present in anaerobic microbial consortium (p. 35)

Fig. 4.4: Prototype of double toilet biodigester (p. 36)

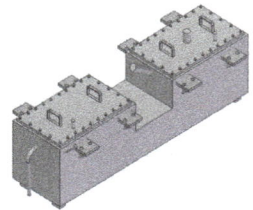

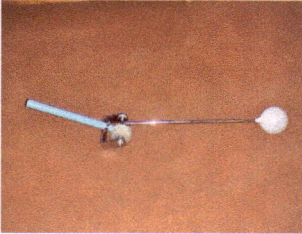

Fig. 4.5: Chlorine ball dispensing assembly (p. 36)

Fig. 4.6: Double toilet biodigester installed under temporary toilet structure for testing under stationary condition; (A): View of biodigester under the toilet shelter, (B): View of toilet installed with biodigester (p. 37)

Fig. 4.7: Testing of siphon system of biodigester in a moving truck (p. 37)

Fig. 4.8: Single toilet biodigester fitted beneath each toilet under the coach (p. 38)

Fig. 4.9: Double toilet biodigester connected with both toilets (p. 38)

Fig. 4.10: Chlorination assembly for dispensing liquid chlorine in the biodigester (p. 39)

Fig. 4.11: Biodigester filled with non-biodegradable material as revealed after opening (p. 39)

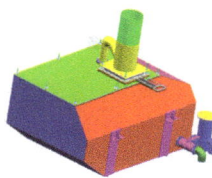

Fig. 4.12: PLC based flapper valve controlled biodigester (p. 41)

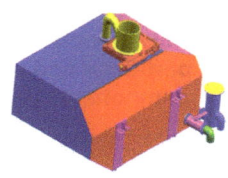

Fig. 4.13: Three dimensional diagram of Manual slider valve based biodigester. (p. 41)

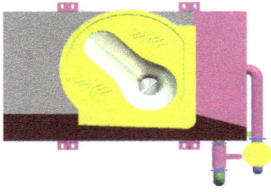

Fig. 4.14: Biodigester with reduced opening at inlet to restrict the entry of bottles, etc. (p. 42)

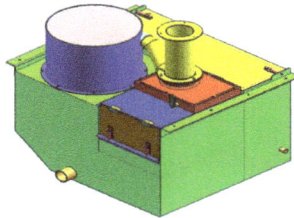

Fig. 4.15: Biodigester with solid liquid separation arrangement. (p. 42)

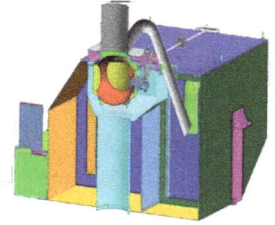

Fig. 4.16: Biodigester with ball valve arrangement for segregation of non-biodegradable waste (p. 43)

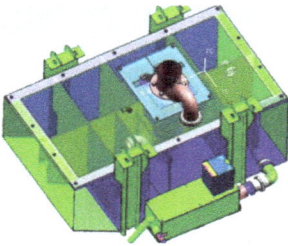

Fig. 5.1: Current biodigester operational in Indian Railways (p. 45)

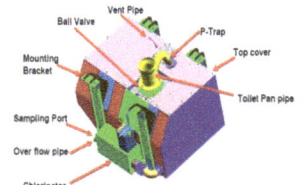

Fig. 5.2: Biodigester showing parts and their positions (p. 46)

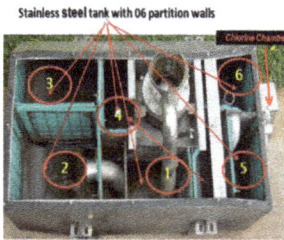

Fig. 5.3: Biodigester showing details of its partitions and other components (p. 46)

Fig. 5.4: Comparative view of filth on passenger coaches with biodigester (A) and without biodigester (B) (p. 49)

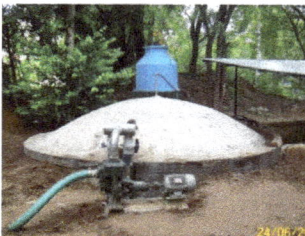

Fig. 5.5: Underground anaerobic microbial inoculum (AMI) generation plant (70 m³ & 14 m³) at DRDE Gwalior (p. 50)

www.ingramcontent.com/pod-product-compliance
Lightning Source LLC
Chambersburg PA
CBHW050519190326
41458CB00005B/1590